THE COMPUTER IN THE SCIENCE CURRICULUM

Janet J. Woerner
Indiana University Northwest

Robert H. Rivers
Purdue University Calumet

Edward L. Vockell
Purdue University Calumet

McGRAW-HILL PUBLISHING COMPANY

New York St. Louis San Francisco Auckland Bogotá
Caracas Hamburg Lisbon London Madrid Mexico
Milan Montreal New Delhi Oklahoma City Paris
San Juan São Paulo Singapore Sydney Tokyo Toronto

 Mitchell Publishing, Inc.
INNOVATORS IN COMPUTER EDUCATION

55 PENNY LANE, SUITE 103 WATSONVILLE, CA 95076
(800)435-2665   IN CALIFORNIA (408)724-0195

Mitchell **McGraw-Hill**
55 Penny Lane
Watsonville, CA 95076

THE COMPUTER IN THE SCIENCE CURRICULUM

1 2 3 4 5 6 7 8 9 0 DOH DOH 9 0 9 8 7 6 5 4 3 2 1

ISBN 0-07-067533-3

The sponsoring editor was James Hill.
The production supervisor was Pat Moran.
The cover design was by Juan Vargas.
The production management was supervised by Richard Lynch, Bookman Productions.
R. R. Donnelley & Sons Company was the printer and binder.

Library of Congress Catalog Card No. 89-062102

CONTENTS

CHAPTER 6

PROBLEM SOLVING AND THE SCIENTIFIC METHOD

149

CHAPTER 7

USING THE COMPUTER AS A TOOL IN SCIENCE

177

PREFACE

This book presents an introduction to the use of the microcomputer to promote science instruction. It assumes no initial knowledge of computers. It does assume some knowledge of theories of science education, but it also summarizes those theories as necessary to give readers the perspective to properly understand this book. This book will be useful to middle school and high school science teachers, to elementary teachers who teach science, to curriculum supervisors, and to others who are interested in integrating the computer into the science curriculum in order to make science instruction more effective.

One of the most "negative" comments we received from reviewers of this book stated: "This all sounds wonderful. If only science could really be taught like this!" We interpreted this reviewer as suggesting that maybe we should "get real"—that we should be aware that most science teachers are going to continue having students memorize definitions so that they can pass tests. In response to such criticism, we have tried to strike a balance between idealism and pragmatism. The fact is that, as we point out in Chapter 1, nearly all major theorists and all major science organizations advocate the use of the learning cycle and inquiry-based approaches that we describe in this book. At the same time, we realize that practical necessity sometimes keeps teachers from engaging in creative, stimulating activities and that competent teachers sometimes employ more "traditional" methods. Our belief is that computerized instruction can be useful to teachers under either set of circumstances; thus, we think this book will be useful to any science teacher who has access to and is willing to use computers for instruction. However, computers themselves do not guarantee good science education. Effective software must be integrated with effective teaching to produce effective learning.

THE SERIES

The Computer in the Science Curriculum is a companion volume to *The Computer in the Classroom* by Edward Vockell and Eileen Schwartz. That book can be considered Module 1 in a series on integrating the computer into the curriculum. Designed for educators in general, it contains a discussion of educational theory applied to the computer across all areas of the curriculum. It also discusses guidelines for software and hardware selection and some programs (such as test generators and gradebooks) that are of interest to teachers in numerous curriculum areas.

This volume is designed for teachers who teach science and for curriculum directors and others interested in applying the computer to their science curriculum. Although it can be regarded as Module 2 in the series, it is also designed to stand alone or to be used as a supplement to other textbooks. For example, many readers will probably use this book as a supplement in a science methods, elementary curriculum, or secondary curriculum course.

In many cases, information that is presented in greater detail in *The Computer in the Classroom* is summarized at appropriate places in the present book, and a citation is given to the chapter in the other book where readers can find more information. In addition, some basic information (such as general guidelines for evaluating drills and a discussion of major brands of computers) is not repeated in this book. Readers can decide for themselves whether they wish to pursue more detailed information in *The Computer in the Classroom* or in some similar general, introductory book on instructional applications of microcomputers.

Other titles that serve as Module 2 in this series include the following:

- *The Computer and Higher-Order Thinking Skills*
- *The Computer in the English Curriculum*
- *The Computer in the Foreign Language Curriculum*
- *The Computer in the Language Arts Curriculum*
- *The Computer in the Mathematics Curriculum*
- *The Computer in the Reading Curriculum*
- *The Computer in the Social Science Curriculum*

Although each of these books is designed as a companion volume to *The Computer in the Classroom*, it is not essential that you read the entire series in order to benefit from one of them. Each book presents a complete and useful set of information to introduce educators to instructional applications of microcomputers.

THE STRUCTURE OF THIS BOOK

Science teachers are often very narrowly defined specialists. A high school chemistry teacher may not be interested at all in elementary school science or in middle school biology. For this reason, we have tried to write this book in such a way as to enable readers to find the information that is uniquely relevant to their personal interests. Nevertheless, we encourage readers to examine the entire text in order to get a good impression of all the possibilities of the computer in science education. One of the marks of a good program or of an effective strategy is that it can often be used at more than one level of education.

Chapter 1 summarizes many of the major issues in contemporary science education and indicates general ways in which the computer can contribute to science education. Chapter 2 examines the instructional theories that relate the computer to effective instruction in science education. Chapter 3 focuses on science education at the elementary school level, Chapter 4 at the middle school level, and Chapter 5 at the secondary level. Chapter 6 discusses the use of the computer to teach scientific thinking skills that can be transferred to many areas of science as well as to areas outside the science classroom. Chapter 7 focuses on tool applications of the computer—including spreadsheets, databases, word processors, and graphic packages—and describes ways in which these can be incorporated into science instruction. Chapter 8 discusses the concept of the microcomputer-based laboratory, in which computerized interfacing devices can make possible data collection and experimentation that would often be impossible or at least impractical with traditional methods. Chapter 9 describes important principles for selecting science courseware and for using computerized materials effectively within the classroom. Chapter 10 ventures into the future and predicts ways in which expected developments in computer technology are likely to have an impact on the science curriculum. Chapter 11 provides summaries of fifty exemplary science programs.

Appendix A includes a glossary of terms, focusing on technical terms from either science education or instructional computing, which need to be clearly defined in order to effectively discuss the application of the computer to the science curriculum. Appendix B includes very short reviews of a much larger number of programs than could be included in the more detailed reviews of Chapter 11. Appendix C provides an annotated bibliography, briefly describing the contents of several books and articles that shed light on current developments in computer applications to science education. Appendix D lists vendors who sell software related to the science curriculum. Finally, Appendix E provides a list of a large number of laserdiscs appropriate for the science curriculum, some information about them, and where to obtain them.

CHAPTER *1*

SCIENCE TEACHING AND THE COMPUTER

IT IS AN ORDINARY third-grade classroom. Student desks are arranged in two groups on either side of the room. The table along the window sill has several different kinds of plants, an aquarium, and a gerbil cage. The bulletin boards, covered with student work, include drawings of constellations. Along one side of the room are a sink, two tables, a cart, and the teacher's desk. One table is strewn with rocks of various types. Various boxes and trays containing batteries, bulbs, wire, nails, and other science supplies are on the other. Lying on one tray is a battery, which has a wire connecting its two poles.

The students are working in groups of two and three. Each group has a beaker in which several rocks have been covered with water; the students are observing the rocks. The teacher intends to pour small amounts of vinegar into the water in a few minutes. After most students describe their rocks in the water, two students get up and carefully carry their beaker to the cart next to the teacher's desk. After inserting a disk into the computer and turning it on, they attach a black plug to the back of the computer, insert a metal cylinder into their beaker of water, and watch a display on the monitor. The students move the cylinder around in the water and rocks, discussing the phenomena they are observing. The teacher walks over to the two and asks them what they are doing. One of the students says, "We're taking the temperature of the water above and below the rocks." "Why would you want to do that?" asks the teacher. "Oh," says the other student, "when you're outdoors on a hot summer day, you can walk into a cave, and it is very cool. I just want to find out if it's cooler under this pile of rocks."

In this third-grade classroom, the computer has become just one more tool for students to use in their inquiry into how the world operates. They use the computer in science, language arts, mathematics. They use the computer as part of their daily class work, and they use it on their own when they have the opportunity.

USES OF MICRO-COMPUTERS IN U.S. EDUCATION

In March 1987, U.S. schools were estimated to have approximately 2.03 million computers available to students in grades 1–12 (Goodspeed, 1988). Today, most U.S. elementary schools have five or more computers, and half of the secondary schools have fifteen or more computers. According to Goodspeed's report, 26 percent of the teachers responded that they use the computer as a supplementary method of improving students' basic skills. Other uses for the computer include teaching computer literacy (19 percent), using it as an effective learning tool (13 percent), and teaching programming (12 percent). Elementary schools use their computers most frequently as a supplementary method for improving basic skills, followed by drill and practice. In junior and senior high schools, computer literacy is the main use, followed by the teaching of programming. In subject-matter areas, computers are used most frequently in mathematics, reading, language arts, and science.

Despite recent developments and improvements in microcomputers and their accompanying software that have made their use more attractive to teachers at both the elementary and secondary levels, some educators are still reluctant to integrate microcomputers into the curriculum. The scenario presented at the beginning of this chapter, while taken from an actual classroom, is not yet representative of most science classrooms.

TRAINING AND PREPARATION OF SCIENCE TEACHERS

Many colleges and universities that train science teachers do not require a separate course in computer applications in education. Pre-service science teachers may take a computer literacy course as an elective, or they may be exposed briefly to microcomputers in their teaching methods courses. Because a microcomputer is only one of many instructional tools that a prospective teacher must study, little time is usually allotted to actually examining and implementing available computer programs. As a result, beginning teachers may lack the expertise required to make use of the technology available to them. Furthermore, many experienced teachers recall the way teaching machines were introduced as a great educational advance in the 1960s and then failed to meet the advertised expectations. Factors like these have made many teachers slow to accept microcomputers in their classrooms.

AIMS OF SCIENCE EDUCATION

In 1982 the National Science Teachers Association (NSTA) issued a position paper on science, technology, and society. In this paper, the NSTA board of directors proclaimed, "The goal of science education during the 1980s is to develop scientifically literate individuals who understand how science, technology, and society influence one another and who are able to use this knowledge in their everyday decision-making. The scientifically literate person has a substantial knowledge base of facts, concepts, conceptual networks, and process skills which enable the individual to continue to learn and think logically. This individual both appreciates the value of science and technology in society and understands their limitations" (1982, p. 1). The position paper described the following critical program emphases that should be considered when developing science curriculum at any level:

1. A minimum time on science learning. The recommendations were:
   • One and one-half hours a week for grades K–3.
   • Two and one-half hours a week for grades 4–6.
   • One hour a day for two full years for grades 7–9.
   • One hour a day for two full years for grades 10–12.
2. Emphasis on programs for all students. Science programs at all levels (K–12) should provide basic concepts for all students, including students with exceptional interest and talent in science.
3. Emphasis on laboratory and field activities that stress
   • Basic inquiry skills at the elementary school level.
   • The application and extension of inquiry skills as a means of obtaining knowledge and resolving problems at the middle school level.
   • Acquisition of knowledge, problem solving, and decision-making abilities at the high school level.
4. Emphasis on science-related societal issues. The recommendations were:
   • Five percent of science instruction should be directed toward science-related societal issues at the elementary level.
   • Fifteen percent at the junior high/middle school level.
   • Twenty percent at the senior high school level.

Although the computer was not specifically mentioned as one of the delivery mechanisms for accomplishing a program with the above emphases, it is easy to see how the computer can be used to enhance concept development, develop inquiry skills, and explore science-related societal issues. The program OH DEER! (MECC) is an excellent example of a simulation that assists learning in all three areas. In the OH DEER! simulation, middle school students are given fixed economic resources and then are faced with the task of maintaining a deer population in a natural area near a small city. Students not only face economic problems but also must deal with social problems in the community arising from the use of various methods of deer eradication.

As a result of working with the OH DEER! program, students develop an understanding of several important science concepts like carrying capacity, predator–prey relationships, and ecological balance. They also develop skill in formulating hypotheses, collecting data to test hypotheses, and drawing conclusions. Societal issues that can be explored include limits on society's willingness to spend money on conservation and social acceptability of various means by which animal populations are controlled.

Since the release of *A Nation at Risk* in 1983, science education has been under fire for producing students who do not understand the science they have studied. Despite the many reports issued since 1983, the aims or goals of science education have changed little, although there has been some rethinking of the relationships and the emphasis of these goals. Questions in science education have focused not on the intent of precollege science education but rather on how the science instruction has been delivered (teaching strategies) and on which goals should be emphasized.

Beginning with the post-Sputnik elementary science curriculum reform movement, the goals in elementary science have focused on the process skills. While content and scientific attitudes make up an equal part of the elementary science course, the emphasis has been to teach children about science by using their natural curiosity to explore and explain their environment. The National Science Foundation (NSF) "alphabet soup" programs (ESS, SAPA, and SCIS) represent, to one degree or another, that approach. Recent re-evaluation has led to a more integrated teaching of those skills, instead of teaching each skill in absence of others.

In middle and junior high school, an effective approach has been to teach basic concepts in the physical, life, and earth sciences in an inquiry or inductive mode, emphasizing science model-building skills. ISCS and IPS were good examples of this approach. The emphasis has been to teach students about science by "doing" science.

At the high school level, curriculum reform focused primarily on updating and logically organizing content. There was also an emphasis in some programs on inquiry development (BSCS Project and ESCP) and on integrating society and technology into the science curriculum (Harvard Project Physics and Engineering Concepts Curriculum Program). Much of the curriculum development projects, however, were about what scientists do—not what regular citizens do with science. Current ideas emphasize the learning of science by all students, not just the college-bound and not just the science majors. In addition, current theories on the constructivist view of learning assume that meaningful learning occurs as a result of active student engagement during learning activities (Tobin, et al., 1988); current reforms therefore emphasize active participation.

The considerations described in the previous paragraphs have led to changes of emphasis in the goals of science teaching and the refinement of methodology related to those goals. Significant changes, however, have occurred only in isolated pockets supported by business and industry, state and federal agencies, and higher education. In reality, while there has been no serious opposition to shifting the emphasis of goals or to more effective methodologies, few of the programs that implemented these approaches have become institutionalized at elementary, middle, or secondary grade levels. This lack of acceptance is due to a number of problems in the schools: lack of appropriate teacher training, school district or regional support people, school district resources to maintain equipment and materials needed to implement such programs, and institutional commitment (as indicated by both criterion-referenced and standardized testing programs) to the "new set of goals" associated with the programs that grew out of the reform movement.

SCIENCE EDUCATION IN THE 1990s

During the summer of 1988, the NSTA board of directors adopted a set of initiatives for the 1990s. Specific recommendations for federal action, chiefly by the NSF, accompany these science education initiatives. The NSTA identifies what they believed to be the four major problem areas facing science education in 1988 and methods to improve science education for all K–12 students:

1. Teacher preparation and staff development
2. Curriculum development
3. Instructional support
4. Research and dissemination

In the area of curriculum development initiatives, the NSTA recommends that there must be curriculum development that instructs teachers on the appropriate uses of technology in the classroom. Instructional support initiatives include providing appropriate electronic technologies to science teachers at all grade levels, providing funds for adequate facilities, and developing regional science centers that disseminate effective teaching practices, science updates, research opportunities, media, and science equipment and supplies available to teachers. In the area of research, the NSTA identifies research on the appropriate uses of technology as a priority for the 1990s.

STAGES OF IMPLEMENTATION

Tinker (1987) identifies three distinct historical phases to the integration of computers into the educational setting. The first is marked by programming, primarily in BASIC, as the only use of the computer in education. While programming has been significantly de-emphasized in science teaching, there continues to be a valid place for students to use high-level languages like Logo to solve problems in science and other curriculum areas.

The second stage involves the use of specific computer-assisted instruction applications. Many of the early programs were simple drills or simplistic simulations. There is no doubt that many of the early application programs were not carefully designed. Goodspeed's (1988) report indicates that teachers are now expressing an interest in having software publishers develop more innovative programs that teach problem-solving and higher-order thinking skills. A number have appeared on the market, and many of these programs are described by Vockell and van Deusen (1989). The ELEMENTARY/MIDDLE SCHOOL SCIENCE INQUIRY SERIES (MECC) is a good example of such programs. These simulation packages include teacher guide materials needed to carefully integrate the simulation or problem-solving situation into the science program.

The third stage, using the computer as a general tool, has gained a great deal of momentum recently. Computerized tools such as word processors, databases, and spreadsheets are more frequently finding their way into science education. These tools not only increase teacher and student productivity but also are analytic tools to ask "what if" questions of all kinds of science phenomena. For example, the organization and analysis of scientific data in spreadsheets that allow the graphing of such data has become a valuable tool in physical science courses at the secondary level.

Other tools, such as information acquisition tools like telecommunications programs and videodisc databases, have found several uses in science teaching. LIFE CYCLE VIDEODISC (Videodiscovery), KIDNET (National Geographic), SCIENCE HELPER K–8 (SCIS), JASON (NSTA/NSF) are examples of these uses. For many years, science teachers have been developing methods to have the computer actually gather, store, manipulate, and graph data from science experiments through various interfacing devices. These microcomputer-based laboratories (MBLs) are becoming more widespread and simpler to use in the classroom.

Tools that allow statistical and mathematical transformations to be made on the data while they are being displayed are currently considered too technical for general classroom use (Tinker, 1987). However, software developments are likely to change this perception. Some programs which have begun to use the computer in this way are EXPERIMENTS IN CHEMISTRY (HRM Software) and INTERFACING COLORIMETRY PROGRAM (Kemtec), both MBL packages. Creativity tools for modeling, imaging, and computer-aided design are also slowly becoming available for the classroom science teacher.

THE COMPUTER AS CHANGE AGENT IN SCIENCE EDUCATION

The power and capabilities of the current generation of microcomputers and the power and capabilities of the computers yet to arrive in the schools should require, as Tinker (1987) puts it, "a radical new approach to instruction throughout the entire curriculum. . . . As the hardware and the software technology matures, there are new possibilities for broad-scale, technologically enhanced changes in the mathematics and science curriculum" (1987, p. 466).

For the most part, the computer is currently supplementing the "old" science curriculum by using "old" teaching strategies through new technology. If the old curriculum and strategies are effective, this presents no problem; but sometimes teachers miss opportunities by adhering to the old ways. For example, two of us are co-authors of a program called BALANCE: PREDATOR–PREY SIMULATION (Diversified Educational Enterprises), which is accompanied by detailed guidelines for using it as part of an inductive process to master important scientific skills. One of us met a high school biology teacher who was using the program in his classroom. After the teacher praised the software for its quality, the author asked how BALANCE was being used in the classroom. The teacher said that he was using the program after a couple of lectures on predator–prey relationships to reinforce prior learning. When asked why he did not use the simulation first (in an inductive or inquiry mode) to help students conceptualize an unfamiliar relationship, the teacher said that while that would be nice, he couldn't afford the time. When the author responded that important scientific problem-solving skills were taught when students went through the simulation inductively, the teacher said that "most students would not be interested in spending the time learning such skills and those skills would not be important to the great majority of the students in my class." As one can see, the power and capability of the computer to simulate a living laboratory has not brought about a major change in the goals or methodology actually pursued in this science classroom. The computer program probably made it possible for this teacher to do his job more easily—and this is a worthwhile goal; but it is obvious that more appropriate use could achieve even more important goals.

This book deals with how the computer can enhance the science teaching and learning process by affecting and transforming the curriculum. Papert (1980) believes that computers can enhance thinking and change patterns of access to knowledge. His belief, which builds on the theory of Jean Piaget, is that the model of successful learning is one that allows children to construct ideas for themselves, using strategies they devise themselves. In many ways, Papert's ideas parallel those that have been at the forefront of science education for many years. The major task for the science teacher in this kind of instructional setting is to provide the appropriate context for learners to construct their own knowledge and develop skill in problem solving, while interacting with the teacher and other students. For example, the teacher may present a problem to the students, provide guidance in solving the problem by giving them a whole-class demonstration or by leading a discussion of the processes involved, and give the computer and software to small groups of students, letting them manipulate the software and hardware to devise a solution. At that point, the ball is in the students' court. With strategic help and encouragement from the teacher, students begin to work to create new skills and concepts while assimilating and accommodating those new skills and concepts into their own current cognitive framework. The students' conception of science and attitudes toward science as a human endeavor are, in the process, transformed.

SUMMARY

If we let it, the computer can increase the effective academic learning time students spend learning the necessary and traditional content goals of the science curriculum. But more than that, the computer can indeed help us transform the emphasis in the present curriculum—learning facts, vocabulary, terms, and sometimes misconceptions or thin conceptions—into an exciting and relevant curriculum that emphasizes concrete, real-world problem solving that has a technological and societal context. In addition, the negative attitudes about science that often accompany the old approach can be changed into positive attitudes through the new approach. The purpose of this book is to help you use computers so that they accomplish both tasks: (1) making teaching and learning of traditional tasks more efficient, and (2) helping to change the emphasis of the curriculum toward goals that need to be accomplished if we are to produce scientifically and technologically literate citizens as well as scientists.

REFERENCES

Goodspeed, J. "Two Million Microcomputers Now Used in U.S. Schools." *Electronic Learning* 7(no. 8) (May/June 1988):16.

NSTA. National Science Teachers Association Position Statement. "Science–Technology–Society: Science Education for the 1980's." NSTA, 1742 Connecticut Avenue, N.W., Washington, D.C., 20009.

Papert, S. *Mindstorms: Children, Computers, and Powerful Ideas.* New York: Basic Books, 1980.

Tinker, R. F. "Educational Technology and the Future of Science Education." *School Science and Mathematics* 87 (1987):466–476.

Tobin, K., M. Espinet, S. E. Byrd, and D. Adams. "Alternative Perspectives of Effective Science Teaching." *Science Education* 72 (1988):433–451.

Vockell, E. L., and R. van Deusen. *The Computer and Higher-Order Thinking Skills.* Watsonville, Calif.: Mitchell, 1989.

CHAPTER 2

STRATEGIES FOR TEACHING SCIENCE

WHAT'S SO HARD ABOUT learning science? Science is important, and it should be fun—so why aren't the goals of science met more easily? The National Assessment of Education Progress (NAEP) reports that the United States is far behind other countries in science learning as measured on standardized tests each year. In addition, Dashiell (1983) reports that about half of all students don't like science by the end of the third grade.

If you ask most students, whether at the elementary or secondary school level, to tell you about science, they will reply that "science is boring" or "all we do is sit and read and write out the vocabulary words and answer the questions at the end of the chapter." Reports indicate that nine out of ten science teachers use a textbook alone 95 percent of the time (NAEP, 1978) and that only 1 percent of the teacher talk in a typical science classroom invites open response. Current high school textbooks can introduce as many as 2,500 new technical terms; a high school foreign language text presents fewer than one-half that number. Coverage of chapters often takes precedence over student mastery of key concepts and scientific problem-solving skills. Few systematic attempts are made to teach centrally important intellectual skills such as reasoning, problem solving, or independent learning.

One of the reasons students find science boring is because teachers often do not convey the "fun" and the "joy" of the a-ha! experience. This excitement is not conveyed, first of all, because *coverage of material* is so important. Using an inquiry mode with hands-on materials and laboratory experiences is very time-consuming, and less material can be covered when those methods are used. In addition, some teachers, particularly those at the elementary level, feel inadequately prepared to teach science. When they teach science at all, they teach it as a reading-only or reading/demonstration activity, in which definitions of words have a connection only to the terms printed on the paper—never to the real world outside the classroom. A third reason for this lack of excitement is "tradition." Despite the tremendous gains made in the past fifteen years in understanding how learning occurs, many teachers continue to teach according to their mental picture of how they themselves were taught in school.

PROBLEMS IN LEARNING SCIENCE

One problem of science learning is that most of us—children and adults—learn most new material in a concrete mode. If we cannot see the thing work, touch it, and fool with it, we cannot make sense of how the thing works. If we do not have any experience with a phenomenon, we cannot think about how it might work; and we may not remember what someone else may tell us about it. This emphasis on the value of concrete experience is the basis for the cognitive development theory formulated by Jean Piaget. Although students become "formal operational" during adolescence and become capable of handling abstractions, abstract reasoning about scientific phenomena will not occur unless it has a concrete foundation on which to build.

It is, of course, necessary to recall specific, factual information in order to engage in scientific thought. Memorized recall of some information in science, such as the symbols for the elements in chemistry class, is necessary if one is to function easily and quickly throughout the year. The need for recall, however, should not be equated with rote memorization. Educational psychologists say that meaningfulness (usually based on concrete experience) provides the basis for useful recall of information. Approaches in which definitions are presented as word-for-word sentences to be memorized and repeated word-for-word on a test lead to rapid forgetting. Operational definitions, those definitions formulated by students and teachers as they observe and manipulate materials and phenomena, are remembered longer because they mean something to those who developed them.

A further problem that science educators have recently recognized is the phenomenon of "children's science"—also called "alternative frameworks" or "misconceptions"—consisting of ideas that children have about how nature behaves. They bring these ideas or alternative conceptions to the science class in a form that is highly resistant to change (Anderson, 1987). Strategies for dealing with misconceptions and prerequisite knowledge are discussed later in this chapter.

USE OF THE LABORATORY

A recent criticism of the U.S. secondary science scope and sequence by the National Science Teachers Association (NSTA) (Aldridge, 1989) calls for beginning the curriculum sequence with concrete, phenomenological, and descriptive approaches early, with semiquantitative and empirical approaches in the middle years and a more theoretical and abstract approach in the later years. This places laboratory activities squarely in the mainstream of the curriculum at the elementary, middle, and most of the high school curriculum. In reality, however, the laboratory often fails to play the role envisioned by the NSTA. Textbook companies have removed much of the laboratory emphasis from many of their textbooks, and many teachers (especially those not trained as scientists) have moved away from using the laboratory in their classrooms. Money, time, and the safety hazards are reasons given for this change.

Sund and Trowbridge (1967) list six kinds of important skills that laboratory work develops in students:

1. Acquisition skills such as listening, observing, investigating, gathering data, and researching
2. Organization skills such as recording, classifying, organizing, outlining, and reviewing
3. Creative skills such as planning ahead, designing a new problem or approach, inventing, and synthesizing
4. Manipulative skills such as using and caring for instruments, demonstrating, constructing, and calibrating
5. Communicative skills such as asking questions, discussing, explaining, reporting, writing, and graphing
6. Safety skills such as handling equipment, hazardous chemicals, and dangerous situations

The computer can provide a viable means of enhancing many of these skills through simulated experiences that fit more easily into the school classroom. Explaining, reporting, and graphing can be done on a computer, using APPLEWORKS and other application programs. Microcomputer-based laboratories can be tied directly into the experiments, and computerized laboratory instruments can collect data and graph it in real time while students conduct and observe the experiment.

APPROACHES TO SCIENCE INSTRUCTION

Science education goals today are a fusion of 1960s and 1980s goals (McIntosh and Zeidler, 1988). These goals give equal emphasis to training future scientists and engineers and to providing scientific training for all students. Most science teachers believe that students should be exposed to knowledge and structure of the disciplines of science, but students should also learn facts, concepts, and principles that would be relevant to the solution of social and technological problems. In addition, some time should be given to the separate disciplines (biology, chemistry, earth science, physics, etc.); but time also needs to be spent showing the interrelationships and the interactions among the different disciplines and between science and nonscience disciplines. The NSTA task force on restructuring the secondary curriculum advocates abandoning the present "layer cake" of courses in biology, chemistry, physics, and sometimes earth science (taken in that sequence), which makes it difficult or impossible for students to see interactions and relationships. The NSTA task force recommends instead that students should study science problems that cut across science disciplines as well as social and technological fields. Finally, teachers also believe that science education should emphasize decision-making skills that demand divergent thought processes and should seek to examine interrelationships between and among environmental systems.

HANDS-ON/ MINDS-ON SCIENCE

The idea that most students need hands-on laboratory experiences to learn concepts is an old one in science education. It is based on the cognitive levels of development originally outlined by Piaget and an ancient proverb: "I hear and I forget. I see and I remember. I do and I understand." Among the current issues in science education is that of "hands-on/minds-on" science. This is similar to the 1960s goal of more concrete experiences but emphasizes the fact that "hands on" can become just as much rote learning as memorizing vocabulary. It is necessary for the teacher to engage the *minds* of students in doing the activities in order to help them learn science.

Hands-on/minds-on science acknowledges that some amount of uncertainty stimulates curiosity and problem-solving behavior. It also acknowledges that too much uncertainty generates anxiety. Each student has a different threshold of uncertainty, which is affected by the amount of learning success and failure the student has encountered.

This idea fits in with current theories acknowledging that learners construct their own knowledge from their experiences. Learners will learn more when incoming bits of information relate substantially to concepts they are building (Andersen, 1986). If the incoming bits of information relate to students' current mental models, they will make sense to students. If they do not relate, then they will either be memorized by rote or forgotten.

Hands-on/minds-on instruction has four main characteristics:

- Provides learning success.
- Contains optimal uncertainty.
- Begins where the student is.
- Contains substantially relatable bits of information.

In curriculum development, these kinds of experiences have been focused around groups of skills, called the process skills. Today's curriculum development in science education has begun to group those skills into the more encompassing categories of description, explanation, prediction, and control. These four functions of science form the focus of current science curriculum design.

Instruction in science must encourage students to be both active and disciplined. Many science programs are dominated by trivial tasks that do not help students develop a coherent scheme. Teachers need to understand both the intended concept (the scientific concept) and the misconceptions held by their students. Teachers need to be involved in more than just presenting material. They are involved in a complex process of modeling how to reason out an event, coaching students through their thinking, and then phasing out their prompts as the students take responsibility for constructing their model of cause and effect.

Ault (1986a) offers some practical methods for achieving these goals, including responding to the spontaneous and changing interests of children, promoting wonder in the natural world, teaching reasoning strategies in response to puzzling experiences, and attending to children's difficult questions about simple events. In intermediate science, he recommends that teachers carefully attend to the long-term development of major conceptions and the interrelatedness of their parts, that teachers provide frequent opportunities to work on problems tailored to the students, that teachers deliberately teach how to think about procedures as well as content, and that they invite students to evaluate their own concept changes and investigative skills (Ault, 1986b).

THE ROLE OF THE COMPUTER

By helping overcome many of the problems discussed in the previous paragraphs, the computer can enrich the science curriculum. By eliminating some of the trivial and boring aspects of the traditional science class, the computer can help put the fun back in science. By packaging science activities in such a way that requires and stimulates student interaction, the computer can permit even relatively nonscientifically trained teachers to lead students in achieving important insights rather than mere rote memorization of information printed in a book. By simulating experiences that would otherwise be, for example, too dangerous, expensive, or time consuming, the computer can help provide an exciting laboratory environment without the frustrations often associated with such labs.

We do not envision the computer as a panacea that will solve all problems of science education. Teachers still need training in science education, and teachers still need to conduct many noncomputerized activities. The thesis of this book, however, is that very often the computer enhances academic learning time and facilitates the application of effective principles of educational psychology. The rest of this chapter specifies exactly *how* the computer can enhance instruction in the science classroom. It is important for teachers to understand these underlying principles. The remaining chapters of this book provide specific examples of computer applications to various areas of the science curriculum.

ACADEMIC LEARNING TIME

Academic learning time (ALT) is defined as the amount of time a student spends attending to relevant and worthwhile academic tasks while performing those tasks with a high rate of success (Berliner, 1984; Caldwell et al., 1982). In any designated subject area, ALT is likely to be more strongly related to academic success than any other variable over which the teacher can exercise control. Vockell and Schwartz (1988) discuss in detail the concept of ALT and its relation to instructional computing. ALT will be discussed here as it relates to science education.

Even if there were no research to verify it, most teachers would assume that students learn more about any given topic or skill when they spend more time on the task. Research has proven that this relationship exists for many academic activities, including science instruction. Simply assigning more study time to science, however, will not automatically increase the student's learning of the topic. The relationship is a bit more complicated than that. For example, not all the time officially scheduled for science is likely to be allocated to that activity. If an hour is assigned to working in a laboratory, but the teacher devotes five minutes at the beginning of the session to returning papers, ten minutes to setting up the lab equipment, and fifteen minutes at the end to cleaning up and making announcements, then only thirty minutes have been allocated to working on experiments.

Scheduled time for instruction merely sets an upper limit on *allocated time*. Similarly, *allocated time* for a subject merely sets the upper limit to *engaged time*, which refers to the amount of time students actively attend to the subject matter under consideration. Finally, even when they are actively engaged in studying science, students learn effectively only when they are performing mental activities at a high rate of success. This smaller amount of time during which students are actively performing at a high rate of success is the factor that is most strongly related to the amount of learning that takes place. In the science classroom, students learn efficiently to the extent that they turn their class and study time into ALT.

Neither "class time" nor "study time" automatically qualifies as ALT, but both may become ALT to the extent that the learner actively attends to relevant tasks with a high rate of success. A student who devotes 100 hours to ALT in a course will learn more than an equally capable student who devotes only 50 hours. However, a person allocating only 50 hours to study and spending 90 percent of it in active academic learning will learn more than an equally capable student who allocates 100 hours but spends only 30 percent of it in active academic learning. ALT is a critical factor no matter which of the general theories of science instruction is embraced by the teacher or school.

ALT can be increased not only by lengthening the school year or school day but also by enabling the teacher to manage a classroom more efficiently and by enabling students to study more efficiently and at a higher rate of success. Although computers cannot lengthen the amount of time spent at school, they *can* help teachers and students perform their tasks more efficiently and with a higher rate of success. In addition, students may simply be motivated to spend more time when they use computers for instruction. It is not imperative that science students have access to computers to make effective use of ALT. Good science teachers are good teachers precisely because they help students make efficient use of ALT. However, it is obvious that the computer *can* make an important contribution to ALT in science. Simply stated, in most situations in which computers enhance learning, they do so because they increase effective ALT. When computers fail to improve learning, it is very often because they do not increase ALT. In looking for areas in which computers can contribute positively to science instruction, it is important to look for ways in which they can increase the ALT available for effective use by students.

Computers with good instructional software can enhance ALT for science in three ways: by permitting learners to study specific scientific information; by enabling students to apply scientific skills to real or simulated problems; and by helping students develop basic tools of learning, which they can apply in a wide variety of settings. The following chapters describe specific ways in which the computer can help increase ALT in various areas of the science curriculum.

> (*Note:* If you have interpreted our discussion of ALT as indicating that it is better to have students actively involved for 100 percent of their time in the pursuit of trivia than to have them actively involved only 75 percent of the time in the pursuit of more important goals, then we have failed to communicate a very important concept. This probably occurred because you hold a previous "alternative conception" of ALT and let that override the concept as we described it. We emphatically recommend devoting ALT to significant educational objectives. If you suffer from this misconception, we urge you to re-examine the preceding paragraphs in this section before continuing with this chapter.)

Our belief is that no matter what approach to science a teacher or school employs, students will learn more effectively within that approach to the extent that they engage in a larger amount of effective ALT. The computer can help students and teachers enhance ALT within any approach. This is not to say that we are embracing an indifferent belief that "all approaches and methods are equally good." Our belief is that a learning cycle approach— exploration, concept introduction, and concept application—is the most natural and flexible approach for teaching most science concepts and processes. To facilitate understanding, we will briefly describe this approach. For more detailed information, refer to Lawson and associates (1989).

During the *exploration* phase of the learning cycle, students learn through their actions and reactions in a new situation as they explore new materials and ideas, with guidance from the teacher and the curriculum materials. These new situations should raise questions that students cannot resolve with their accustomed ways of thinking (Lawson et al., 1989). It should also lead to the identification of patterns of regularity in the scientific phenomena being studied.

The second phase, *concept introduction*, includes the introduction of a new concept or set of concepts such as cell, metabolism, kinetic energy, tsunami, and so on. This phase usually involves the introduction of useful terms to describe the patterns identified in the exploration phase. The new concept may be introduced by the teacher, the textbook, a video segment, or a computerized tutorial. This step should follow exploration and relate directly to incorporating the pattern identified during the exploration phase into a usable concept. If the entire pattern has not been identified by students, the teacher should review the concept and reveal the missing portions. It is important that both critical and variable attributes of the concept are dealt with.

In the last phase of the learning cycle, *concept application*, students apply the newly learned concept or thought pattern to additional samples, some of which are examples of the concept and some of which are nonexamples. Students then broaden the range of applicability of the concept and can generalize it to other situations. In addition, application activities aid students whose conceptual reorganization takes place more slowly or students who did not relate the original explanation to their experiences.

We advocate using the computer in such a way as to incorporate this learning cycle approach into the science curriculum. At the same time, we recognize that when lower-level knowledge and comprehension objectives are emphasized in science, it is appropriate to use a diagnostic prescriptive approach with immediate feedback and positive reinforcement. ("Lower level" does not mean trivial or unimportant. The term simply recognizes that some level is at the lowest levels of Bloom's taxonomy.) We also recognize that competent teachers might differ with our emphasis on these approaches; thus, we have written this book to be useful for teachers who base their teaching on other approaches as well.

We therefore do not recommend that the microcomputer should become one more subject area for students to study; rather, it should become a means to facilitate and enhance science teaching and learning. In the science curriculum, the computer can provide a method of integrating science and various thinking skills to provide a more meaningful experience for the student. In addition, the computer can play a vital role because it has the capacity to both motivate learners and focus their attention more effectively on the task at hand.

IMPORTANT PRINCIPLES OF EDUCATIONAL PSYCHOLOGY

Vockell and Schwartz (1988) provide a detailed discussion of several principles of educational psychology that the computer can help incorporate into the classroom to make instruction more effective (Table 2.1). Nearly all these principles are applicable to science instruction. The value of concrete experiences in science has already been discussed earlier. We now focus on five additional principles that are especially relevant to the science curriculum: mastery learning, direct instruction, student misconceptions, prerequisite knowledge and skills, and questions teachers ask. Two other principles (cooperative learning and peer tutoring) are discussed in detail in Chapter 9 of this book. Many of the other principles are also very important for science instruction, as they are for all curriculum areas. For a more detailed discussion of these other principles, readers should refer to Vockell and Schwartz, 1988.

Table 2.1. Summary of major instructional principles and guidelines for using the computer.*

Principle: Mastery Learning
Summary: Given enough time, nearly all learners can master objectives.
Guidelines: 1. Use programs that provide extra help and practice toward reaching objectives.
             2. Use programs to stimulate and enrich students who reach objectives early.
             3. Use record-keeping programs to keep track of student performance.

Principle: Direct Instruction
Summary: If teachers describe objectives and demonstrate exact steps, students can master specific skills more efficiently.
Guidelines: 1. Use programs that specify exact steps and teach them clearly and specifically.
             2. Show the relationship of computer programs to steps in the direct teaching process.

Principle: Overlearning
Summary: To become automatic, skills must be practiced and reinforced beyond the point of initial mastery.
Guidelines: 1. Use computer programs to provide self-paced, individualized practice.
             2. Use computer programs that provide gamelike practice for skills that require much repeated practice.
             3. Use computer programs that provide varied approaches to practicing the same activity.

Principle: Memorization Skills
Summary: Recall of factual information is a useful skill that enhances learning at all levels.
Guidelines: 1. Use computer programs to provide repeated practice and facilitate memorization.
             2. Use programs designed to develop memory skills.

Principle: Peer Tutoring
Summary: Both tutor and pupil can benefit from properly structured peer tutoring.
Guidelines: 1. Have students work in groups at computers.
             2. Use programs that are structured to help tutors provide instruction, prompts, and feedback.
             3. Teach students to give feedback, prompts, and instruction at computers.

Principle: Cooperative Learning
Summary: Helping one another is often more productive than competing for scarce rewards.
Guidelines: 1. Have students work in groups at computers.
             2. Use programs that promote cooperation.
             3. Provide guidelines for cooperative roles at computers.

Principle: Monitor Student Progress
Summary: Close monitoring of student progress enables students, teachers, and parents to identify strengths and weaknesses of learners.
Guidelines: 1. Use programs that have management systems to monitor student progress.
             2. Use record-keeping programs.
             3. Use computer to communicate feedback.

(continued)

Table 2.1 (continued)

Principle: Student Misconceptions
Summary: Identifying misconceptions helps develop an understanding of topics.
Guidelines: 1. Use programs to diagnose misconceptions.
             2. Use programs to teach correct understanding of misunderstood concepts.

Principle: Prerequisite Knowledge and Skills
Summary: Knowledge is usually hierarchical, and low-level skills must be learned before higher-level skills can be mastered.
Guidelines: 1. Use programs to assess prerequisite knowledge and skills.
             2. Use programs to teach missing prerequisite skills.

Principle: Immediate Feedback
Summary: Feedback usually works best if it comes quickly after a response.
Guidelines: 1. Use programs that provide immediate feedback.
             2. Use programs that provide clear corrective feedback.

Principle: Parental Involvement
Summary: Parents should be informed about their children's progress and assist in helping them learn.
Guidelines: 1. Use computers to communicate with parents about educational activities and progress.
             2. Exploit home computers.

Principle: Learning Styles
Summary: Learners vary in preference for modes and styles of learning.
Guidelines: 1. Use programs that appeal to students' preferred learning styles.
             2. Use programs that supplement your weak teaching styles.
             3. Use programs that employ a variety of learning styles.

Principle: Classroom Management
Summary: Effective classroom management provides more time for instruction.
Guidelines: 1. Use the computer as a tool to improve classroom management.
             2. Use programs that have a management component.

Principle: Teacher Questions
Summary: If teachers ask higher-order questions and wait for students to answer, higher-level learning is likely to occur.
Guidelines: 1. Select programs that ask higher-level questions.
             2. Use programs that individualize pace of instruction because wait time is likely to be better than with traditional instruction.

Principle: Study Skills
Summary: Effective study skills can be taught, and these almost always enhance learning.
Guidelines: 1. Teach students to use the computer as a tool to manage and assist learning.
             2. Use programs that teach thinking skills.
             3. Teach generalization of thinking and study skills across subject areas.

Principle: Homework
Summary: When homework is well planned by teachers, completed by students, and related to class, learning improves.
Guidelines: 1. Assign homework for home computers.
             2. Have students do preparatory work off the computer as homework.

Table 2.1 (continued)

Principle: Writing Instruction
Summary:　Writing should be taught as a recursive process of brainstorming, composing, revising, and editing.
Guidelines:　1.　Use word processors for composition.
　　　　　　2.　Use programs that prompt writing skills.
　　　　　　3.　Teach students to use grammar and spelling checkers effectively.

Principle: Early Writing
Summary:　Encourage even very young children to write "stories."
Guidelines:　1.　Use simple word processing programs.
　　　　　　2.　Use programs that combine graphics with writing.
　　　　　　3.　Use graphics programs to stimulate creativity.

Principle: Learning Mathematics
Summary:　Concrete experience helps students understand and master abstract principles.
Guidelines:　1.　Match programs to children's level of cognitive development.
　　　　　　2.　Use programs that provide concrete demonstrations with clear graphics.

Principle: Phonics
Summary:　Instruction in phonics helps students "break the code" and develop generalized word attack skills.
Guideline:　1.　Use programs that combine sound with visual graphics to teach the sight/sound relationships of reading.

Principle: Reading Comprehension
Summary:　Students often learn better if reading lessons are preceded by preparatory materials and followed by questions and activities.
Guidelines:　1.　Use programs that have pre- and postactivities to accompany them.
　　　　　　2.　Use computer programs before or after traditional reading materials.

Principle: Science Experiments
Summary:　Students learn science best if they can do concrete experiments to see science in action.
Guidelines:　1.　Use computer simulations.
　　　　　　2.　Use tutorial and drill programs with concrete graphics.
　　　　　　3.　Use database and word processing programs to manage and report noncomputerized science experiments.
　　　　　　4.　Use science interface equipment to manage and analyze science experiments.

*From Vockell and Schwartz, 1988.

Mastery Learning

The principle of mastery learning states that given enough time and help, about 95 percent of the learners in any group can come to a complete mastery of the designated instructional objectives. Traditional instruction holds time constant and allows achievement to vary within a group. For example, a traditional grading period for science may last nine weeks; at the end of that time, students who have thoroughly mastered the designated objectives receive grades of *A*, those who have mastered very little get grades of *F*, and so on. Mastery learning reverses this relationship by holding achievement constant and letting the time students spend in pursuit of the objectives vary. In the same science program, a few students might meet the standards in twelve weeks; most might meet the standards in nine weeks; but a few students might take twenty or twenty-five weeks to meet these standards.

Mastery learning is not synonymous with pass/fail grading, nor does it imply that "standards should be lowered." When mastery learning is successful, high standards are articulated, and students receive ample time and help to meet these high standards. Additional information about mastery learning can be found in Guskey and Gates (1986), Slavin (1987), and Levine (1987).

Mastery learning has received formal emphasis only in the past twenty years; but, informally, students and teachers have known about this principle for a long time. For example, students having trouble in any subject usually believe they can master it if they are given enough time. Even teachers who work within systems where they must force all students to work at the same pace find that they can teach more effectively when they incorporate some aspects of mastery learning into their instruction.

Two problems often arise with mastery learning: (1) Grouping and scheduling may become difficult. It is easier to force people to work at a constant pace and to complete tasks at a predictable rate than to permit wide variations in activities within a class. (2) While slower learners spend extra time on minimum standards, faster learners may be forced to wait, when they could be progressing to higher-level achievement. These problems are not insurmountable. They are overcome by providing individualized attention, setting high but attainable standards, and making additional materials available for students who master objectives more quickly than other learners.

Computers can aid mastery learning in three ways:

1. Many students need additional time and individualized practice with feedback to meet objectives. Computer programs can often provide

opportunities to study at times and at a pace suited to the individual's needs.

2. Additional programs can be made available for students who master objectives quickly. These additional programs can either provide more intense study of the same objectives, move on to higher objectives, or integrate the objectives covered in the unit with other objectives.

3. Gradebook, record keeping, and other management programs can help teachers keep track of student performance and coordinate instruction.

Mastery learning has for many years worked well without the aid of computers. However, the wise use of computers can often make mastery learning work more effective.

Direct Instruction

Closely related to mastery learning is direct instruction—academically focused, teacher-directed interactions using sequenced instructional materials. Further characteristics include

- Goals that are clear to students.
- Sufficient and continuous time allocated for instruction.
- Introduction of terms and concepts.
- Monitoring of student performance.
- Questions at an appropriate cognitive level, enabling students to produce numerous correct responses.
- Immediate and academically oriented feedback.

Some scientific topics and skills, such as psychomotor skills involved in science, are particularly amenable to direct instruction. Direct instruction has proven especially effective in teaching basic skills (such as how to use a microscope or the definitions of important terms in biology) and skills that are fundamental to more complex activities (such as the steps in a complex analytic procedure in chemistry or mastering cell staining or electrophoresis technique in biology). Direct instruction is not as likely to be useful for teaching less structured units, such as developing scientific creativity or the discussion of social issues in science.

In addition, the application of direct instruction to learning cycles in science may be different from its application to mathematics or language arts instruction. In many cases, effective direct instruction in science is synonymous with what science educators have called "guided discovery."

Direct instruction is important during each phase of the learning cycle. In the exploration and application phases, it is essential that the teacher intervene directly in the activity of students if they are not asking the right question, not making appropriate observations, getting sidetracked into an unprofitable learning behavior, not collecting appropriate data, or not paying attention to the data that they are collecting. In such cases, direct instruction means that the teacher asks questions and provides suggestions at appropriate times to help students focus on profitable activities. Teachers must not let students digress on irrelevant or erroneous activities, and teachers must refrain from giving simplistic, fact-oriented feedback that will promote trivial memorization rather than understanding of the processes and concepts under consideration. In these phases of instruction, improper feedback focusing heavily on "right answers" could cause students to memorize trivia or purely rote information; such outcomes are the exact opposite of what students should accomplish during these phases. Students must be guided by the teacher (or textbook or computer program) to construct or apply the concept themselves. Students should "know" the concept is a good one because it helps them understand a real problem or because it helps them explain the data they have collected, not because "the teacher (or the textbook or the computer) said so." The important thing in both phases is that students interact with materials and work out patterns and concepts for themselves.

Direct instruction is also commonly used during the term introduction, or concept introduction, phase of the learning cycle. Interaction during this phase should be structured, but not authoritarian. Learning best takes place in a convivial atmosphere, where there is a give and take between teacher and students—each contributing something to the invention of the pattern and accompanying operational definition of the term, each suggesting critical and variable attributes related to the concept, and each suggesting ways to apply the concept.

Some educators misinterpret the role of direct teaching in science. They equate direct teaching with the memorization of a term, but this is far from an accurate perception. Direct instruction can and should focus on all phases of the learning cycle, emphasizing the application and practical use of higher-order thinking skills.

Direct instruction is one of the activities that the computer performs especially well. Figure 2.1 provides an example of a program designed to teach middle school or high school students how to make and record scientific measurements. The program states its objective, provides a tutorial upon request, and then provides numerous opportunities for practice, with immediate feedback.

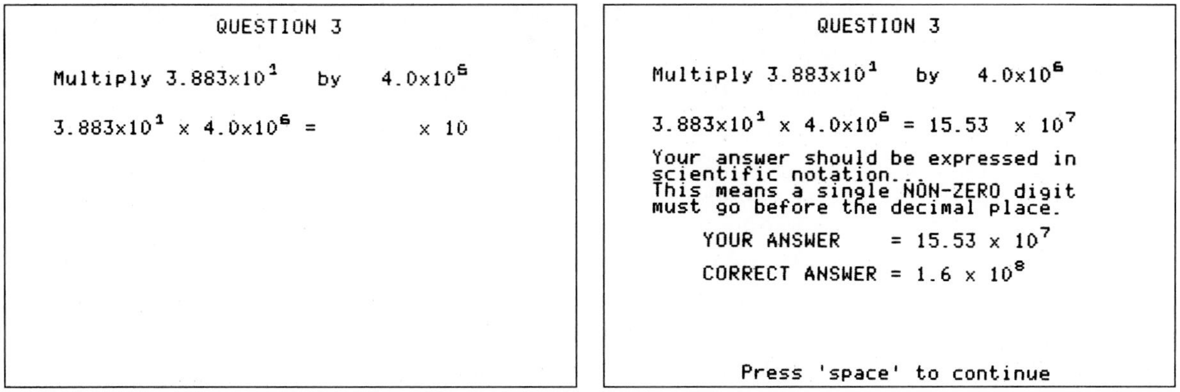

```
┌─────────────────────────────────────┐   ┌─────────────────────────────────────┐
│             QUESTION 3              │   │             QUESTION 3              │
│                                     │   │                                     │
│ Multiply 3.883x10¹   by   4.0x10⁶   │   │ Multiply 3.883x10¹   by   4.0x10⁶   │
│                                     │   │                                     │
│ 3.883x10¹ x 4.0x10⁶ =       x 10    │   │ 3.883x10¹ x 4.0x10⁶ = 15.53  x 10⁷  │
│                                     │   │ Your answer should be expressed in  │
│                                     │   │ scientific notation.                │
│                                     │   │ This means a single NON-ZERO digit  │
│                                     │   │ must go before the decimal place.   │
│                                     │   │    YOUR ANSWER    = 15.53 x 10⁷     │
│                                     │   │    CORRECT ANSWER = 1.6 x 10⁸       │
│                                     │   │                                     │
│                                     │   │       Press 'space' to continue     │
└─────────────────────────────────────┘   └─────────────────────────────────────┘
```

(a)   **(b)**

Figure 2.1. Screens from MATHEMATICS FOR SCIENCE (Merlan Scientific) for teaching scientific notation to middle school or high school students. Each program in this series states its objective, provides a tutorial if requested, and provides numerous opportunities to practice.

It should be clear that even if a program itself does not incorporate all the features of direct instruction, teachers can use almost any good computer program as a component of effective direct instruction. It is merely necessary that the teacher clearly point out the objective of a unit of instruction, supply the computer program, point out the connection between the objective and the computer program, and then monitor the student's use of the program as necessary. It should also be clear, however, that when they are used improperly, computer programs can seriously interfere with effective learning by providing improper direct instruction. For example, a program that gives purely factual feedback during the exploration or application phase (rather than encouraging students to clarify and internalize the concept under consideration) could cause students to restrict rather than expand their opportunities for understanding and applying the concept. Programs that give specific, factual feedback are not necessarily bad programs—they may be ideal programs during the concept introduction phase; even during the exploration and application phases, they can be effective if their use is coordinated by a teacher who integrates them with appropriate teaching strategies. The key point is that teachers must understand the purpose of a particular phase of instruction, understand the capabilities of a specific piece of software, and integrate the appropriate software with the unit of instruction.

The research on mastery learning and direct instruction is discussed in greater detail in Guskey and Gates (1986), Levine (1987), Rosenshine (1986), Slavin (1987), Vockell and Schwartz (1988), and Wang and Walberg (1985).

Student Misconceptions

As mentioned earlier, children often have initial misconceptions regarding scientific concepts, and their misconceptions provide a faulty foundation for learning accurate scientific information (Schoon, 1989). Teachers who wish to teach effectively need to develop activities and lessons that acknowledge these misconceptions and take steps to overcome them. If a teacher knows what misconceptions are held by students, the teacher can often help them overcome these misconceptions and develop a correct understanding of a topic being taught. Teachers do this by confronting students with new situations where their alternative conceptions or misconceptions cannot help them explain the patterns of data they are collecting.

The conceptions about a scientific topic are likely to differ even among the students within a single class. When students have misconceptions about a topic, their status is often worse than if they thought they knew nothing about the subject under consideration. The false information transfers negatively to the instructional situation and makes it even more difficult for the learner to reach instructional objectives. Good teachers identify these misconceptions and help students overcome them, so that learning can proceed effectively.

The role of the computer in dealing with misconceptions is similar to its role in dealing with prerequisite knowledge. The computer can effectively help determine what misconceptions are held by students and then provide experiences to help students confront them.

Prerequisite Knowledge and Skills

Knowledge is often hierarchical. Often the best way to ensure performance on higher-level objectives is to identify the prerequisite skills needed for a current unit of instruction and ascertain that students have mastered these prerequisites. If the teacher performs a task analysis to determine what skills students must perform before learning a more complex skill and then verifies that students already possess all these prerequisites, then students will almost always master the new ability quickly and easily. Students have trouble with new skills mainly because they lack the prerequisite skills.

From the teacher's perspective, learning is ineffective because task analysis or preassessment has not been conducted. If preassessment shows that some students are lacking some prerequisite skills, these should be taught before moving to the targeted skill. Of course, if students not only lack prerequisite skills but also hold misconceptions, then the need for additional experiences is even more serious. The value of organizing instruction hierarchically and of ascertaining that students have mastered prerequisite skills is discussed in detail in Gagné (1985).

The computer often fits perfectly into schemes to assess prerequisite skills prior to a unit of instruction. Two of the main reasons teachers do not perform these preassessments are (1) it takes a great deal of time to diagnose all students, and (2) if students are deficient in prerequisite skills, there is often no good way to help them remediate these deficiencies. The computer can help either by (1) performing the assessment or (2) delivering the remediation, or by doing both.

Figure 2.2 shows a good example of effective computerized diagnosis. This program presents students with scientific problems to solve. It records not only answers but also the types of thinking in which students engaged while solving the problems and the types of mistakes they made. Teachers can later access the results to determine what their students currently know about scientific problem solving.

Even without computerized diagnosis, the computer can make a valuable contribution by helping students develop prerequisite skills before beginning units of instruction. For example, prior to teaching a unit that requires measurement as part of a scientific investigation, a teacher could make available the MATHEMATICS FOR SCIENCE program shown earlier in Figure 2.1. Teachers are often aware that their students lack prerequisites, but they just do not have time to work with students who were supposed to master prerequisites in a previous unit or in a previous course. Computer programs can help solve this problem by providing individualized remediation for subskills as necessary.

A further impediment to teaching prerequisite skills is that it is time consuming for individual teachers to perform task analyses needed to identify them. It is possible to develop computerized databases that provide task analyses for any topics covered in the curriculum. When a student encounters difficulty, it would be possible for the teacher to generate from the database a list of prerequisite skills accompanied by a list of computer programs that would help the student master them. Currently, we do not know that such detailed databases exist. The closest we have seen are programs that store objectives to generate individualized educational programs (IEPs) for special education students and teachers.

Finally, we should look into the future. From the field of artificial

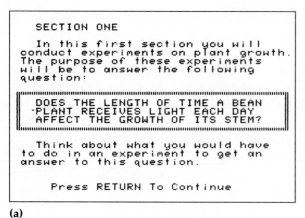

(a)

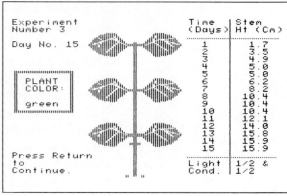

(b)

Figure 2.2. A series of screens from the Illinois Scientific Problem Solving Test (Rivers and Luncsford). This test presents biological problems to solve and tracks the decisions students make to find a solution. Screen (a) shows the presentation of a problem related to plant growth. Screen (b) is a data collection screen. Screen (c) is a question screen, requiring students to interpret the data they have collected.

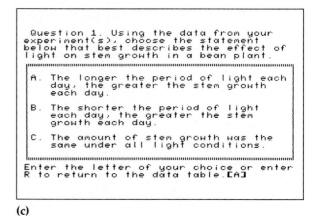

(c)

intelligence, there is emerging a new branch of computerized education known as intelligent computer-assisted instruction (ICAI). Programs employing this strategy will look for patterns in student responses and will automatically branch to remedial units of instruction when students appear to lack the skills needed to accomplish a designated task. At the time of this writing, there are not yet any good examples of ICAI on the market; but MENDEL (described in Chapter 10 of this book) is a good example of an ICAI program in the final stages of development.

Questions Teachers Ask

Student achievement rises when teachers ask questions that require students to apply, analyze, synthesize, and evaluate information in addition to simply recalling facts—and when teachers give students time to think in order to answer these questions. Two factors that seriously inhibit learning are that teachers tend to ask only questions that require rote memorization and that they expect students to answer within about one second after a question is asked. Research shows that when teachers ask higher-order questions and give appropriate feedback for answers, students learn higher-order skills more effectively. In addition, research indicates that if teachers pause a few seconds longer and provide appropriate prompts, students can often answer higher-order questions and benefit accordingly. Excellent summaries of research on teacher questioning and wait time can be found in Barell (1985) and Tobin (1987).

Computers are not automatically superior to teachers at asking higher-order questions that involve nonrote skills. In fact, a very large number of drill programs require nothing more than rote responses. Note, however, that even if all CAI programs required merely rote responses, they could still enhance higher-order thinking by performing this rote function and freeing teachers to engage in more higher-order thinking with students. Of course, it is also possible to program the computer to ask higher-order questions, as Figure 2.3 indicates. If science teachers are interested in promoting nonrote skills, they should examine programs carefully and select software that requires higher-order performance.

Computers do have an inherent advantage over teachers with regard to how long they are willing to wait for the student to respond. When a teacher asks a question, a lengthy silence may signal a coming disruption in the teaching process; and many teachers feel compelled to indicate that the student is "wrong" or to call on another student almost immediately. Actually, the most sensible process in which to engage when asked a

Figure 2.3. A screen from INVISIBLE BUGS (MECC), showing a higher-order question.

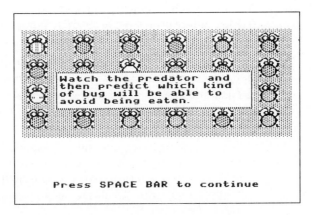

Watch the predator and then predict which kind of bug will be able to avoid being eaten.

Press SPACE BAR to continue

difficult question is to pause, analyze the problem, bring to mind relevant information, develop a tentative answer, check the validity of this tentative answer, and then give the answer out loud. It's difficult to do this within the short time provided by most teachers; and so students (at best) "think on their feet." If they understand the question, they either give a memorized (or previously thought out) response or start talking and develop their answer while they are giving it. This may be a useful strategy for winning prizes on a game show, but it hardly enhances effective scientific thinking. (In fact, students who benefit are probably the other thoughtful students in the rest of the class, who go through all the appropriate steps while the teacher is calling on the first student.) The computer solves this problem simply by waiting as long as necessary for the student to respond. There is no ominous silence after the computer presents a question, and no pressure to move on to another student. When asked a higher-order question, the student can pause, go through the appropriate steps, and then respond.

In addition, the computer can present situations that call for higher-order thinking. By using the inquiry strategies described in Chapter 6, teachers can use effective programs as an opportunity to ask students higher-order questions.

TYPES OF SOFTWARE

The characteristics of good drills, tutorials, simulations, and tool applications of computers are discussed in detail in Vockell and Schwartz (1988). This section focuses on those features that are especially important in science software.

Drills and Tutorials

Drill-and-practice programs give students feedback for responses as they practice skills or concepts previously taught. Practice of a concept or skill beyond initial mastery is essential; but this does not mean that drills are essential. The additional practice can often occur in the context of a practical application or as part of the progression to the next objective in a unit of instruction. However, for many students, a drill on basic information can be a valuable tool for making this information a part of his or her basic repertoire before moving on to higher-level objectives.

The following five guidelines are especially pertinent to science drills:

1. Unsupervised drills are part of the practice stage, not of the learning stage of instruction. Drills are useful only if the student gives the right

answer about 90 to 95 percent of the time. If drills are to be used in the learning stage of instruction, provide them with teacher guidance, as described in Chapter 9.

2. There is no point in drilling on information that the student does not understand. If a student has misconceptions about information, a drill is not likely to correct them. (It *is* possible for students to use negative feedback from a drill as a basis for asking the right questions to overcome misunderstandings; however, they have to *learn* to do this. Drills do not make this happen.)

3. Drills can be overdone. Once students have thoroughly mastered an objective, repeated performance of the same activities that helped attain that mastery is likely to become boring. If repeated practice is necessary (and it often is), it should be presented in an interesting way that is likely to reduce boredom.

4. Drills are often used when they should not be used because they are easy to employ, even by teachers who do not thoroughly understand the concepts, skills, or principles covered by the drill. Select tutorial, simulation, or tool software when your goal is to teach new information. Drills are designed *solely* to provide practice on objectives already taught in some other context.

5. When drills are used inappropriately, they often backfire by giving the impression that an area of science is difficult and abstract and that the best (or only!) way to learn it is through memorization without understanding the concepts.

Despite these caveats, good computerized drills can provide excellent opportunities for practice and feedback with regard to important scientific concepts, principles, and skills. Good drill programs are in game formats, presentation/question formats, or quiz formats and may cover a wide variety of topics. Figure 2.4 shows four examples of good science drills.

A drill-and-practice program must be carefully designed because it not only provides time to master objectives but also has the opportunity to turn students off to science. At their best, drills help students develop automaticity—important skills and concepts become so familiar that students can automatically use them when pursuing higher-level activities. At their worst, drills trivialize science by making students focus on lower-level activities to the exclusion of applying these skills and concepts to the higher-level activities that science is really about. Well-designed science software offers some advantages not displayed by traditional drill activi-

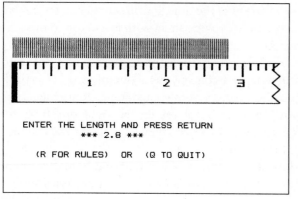

(a)

In the diagram, what is the letter
used to label the angle of reflection?
       TYPE 99 TO SEE DIAGRAM.
       TYPE 999 TO SEE EQUATIONS

What is the answer, JAN?A

   Wrong answer, JAN.
The picture shows a beam of parallel
rays being reflected from a smooth plane
surface. Since all rays are reflected
from the same surface, the rays remain
parallel after reflection.
   The angle of reflection is the angle
between the normal and the reflected
ray.

What is the answer, JAN

(b)

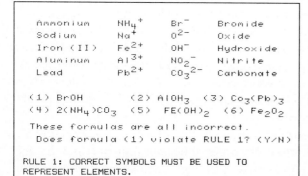

(c)

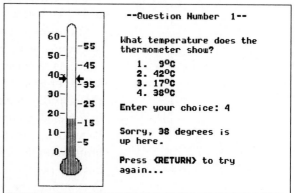

(d)

Figure 2.4. In CHEMISTRY ACCORDING TO ROF (ROF) and SCIENCE SKILL DEVELOPMENT (Edutech), drills are presented in a presentation/question format: Screen (a), SCIENCE SKILL DEVELOPMENT; screen (b), WAVES (J & S Software); screen (c), CHEMICAL FORMULAS (ROF); screen (d), TEMPERATURE EXPERIMENTS (Hartley).

ties (e.g., completing workbook pages, doing problems and questions, writing out the vocabulary).

The major advantage of a computerized science drill is its ability to present the material in a colored, animated, visual way, while encouraging involvement of multiple modalities. This helps students understand diagrams and focuses the learning in the visual mode, while allowing tactile or kinesthetic involvement—and often auditory prompts. In addition, the computer can provide immediate feedback to allow students to correct their mistakes immediately, instead of waiting until the homework is checked the next day.

Tutorials present new information to students and attempt to correct or help students as they proceed through the program. The main difference between a drill and a tutorial is that the tutorial is designed to be used during the learning phase of instruction. It teaches new information. The major advantages of a computerized science tutorial are its ability (1) to employ branching programmed instruction and thereby focus on the exact needs of the individual learner; (2) to present the material in a colored, animated, visual mode, while encouraging involvement of multiple modalities; and (3) to evoke active interaction on the part of the learner.

These advantages are often more imagined than real. There are not many good science tutorials on the market. Most tutorials that we have reviewed are merely electronic page turners that require minimal genuine interaction on the part of the learner and show no real advantage over books or films on the same topic. For example, WEATHER FRONTS (Teach Yourself by Computer) (Figure 2.5) presents accurate information about weather patterns; but a student would probably be better off reading a good book chapter (where it would be possible to look back in the book to review previous concepts) or viewing a good film (where the graphics would be much better). The quality of science tutorials is likely to improve with improvements in interactive video using such programs as HYPERCARD

```
A FRONT IS A BOUNDARY AREA BETWEEN 2
BODIES OF AIR CALLED AIR MASSES. AIR
MASSES ARE USUALLY DEFINED BY MARKEDLY
DIFFERING TEMPERATURES. AS THESE MASSES
COME TOGETHER THEY TEND TO FORM A ZONE
OF RAPIDLY CHANGING TEMPERATURE, RATHER
THAN MIXING. FRONTS ARE INDICATED BY
LINES ON WEATHER MAPS. THIS LINE IS
THE PLACE WHERE THE MASSES MEET AT THE
GROUND. THE AIR MASS SLOPES UP FROM THAT
LINE IN A BOUNDRY 50 TO 500 MILES IN
LENGTH.
            PRESS SPACE BAR TO CONTINUE
```

(a)

```
WEATHER FRONTS          FIG#1
PRESS RETURN TO REVIEW, ELSE SPACE BAR
```

(b)

Figure 2.5. In WEATHER FRONTS, screen (a) presents the general explanation of weather fronts with the accompanying graphic in screen (b).

and LINKWAY and with the application of artificial intelligence to ICAI. These developments are discussed in Chapter 10.

At the present time, tutorials appear primarily as remedial or introductory segments in drill programs. That is, when a student gives a wrong answer in a drill, a tutorial screen or two may appear to help the learner come to the correct understanding of the concept or skill under consideration. Figure 2.6 shows an example of some tutorial segments that appear in computerized drills.

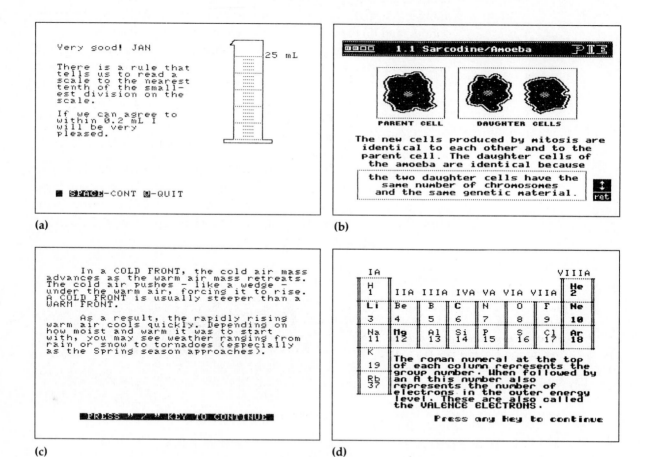

Figure 2.6. Tutorial segments that appear in drill and practice programs: Screen (a), CHEMISTRY ACCORDING TO ROF (ROF); screen (b), REPRODUCTIVE SYSTEM (Prentice-Hall); screen (c), WEATHER FRONTS (Diversified Educational Enterprises); screen (d), ELECTRON PROPERTIES (J & S Software).

Simulations

Computerized simulations make it possible to accomplish activities that would otherwise be too dangerous, expensive, time consuming, unethical, impossible, or impractical to carry out in the science classroom. Simulations give the student the feeling of having participated in an activity (such as an experiment) and offer the educational advantages of participation without requiring the student to physically participate in the actual activity. Because science is inherently concerned with the development of models and because these models can be stated in terms of algorithms and mathematical formulas, simulations for science activities are extremely prevalent. They are most useful in helping students develop concepts that would otherwise be hard to introduce (e.g., food chain) and in teaching higher-order thinking skills such as developing and testing hypotheses. Figure 2.7 shows four examples.

In evaluating science simulations, it is important to consider two major features: (1) the adequacy of the scientific model and the assumptions underlying the model, and (2) the pedagogical materials that surround the model and make it into an effective instructional unit. These pedagogical materials may be part of the program, or they may be separate materials (e.g., print or video) that accompany the software.

Tool Applications

Besides providing drills, tutorials, and simulations, the computer can serve as a simple electronic tool for students and teachers. In this role, the computer performs tasks that students or teachers would normally perform anyway, but it performs these tasks more conveniently. With onerous tasks out of the way, students and teachers can focus on the science concepts, principles, and skills that are really the focus of instruction. Examples of tool applications include programs such as graphing, database, spreadsheet, word processing and interface instruments that collect and tabulate data. These are discussed in detail in Chapters 7 and 8.

SUMMARY

This chapter has described several problems and basic approaches to teaching science and has introduced ideas for integrating the computer into the science curriculum. Computers can be usefully employed within any of the described approaches to science instruction. Subsequent chapters of this book describe specific strategies of integration more completely.

There is an underlying theoretical framework for using the computer

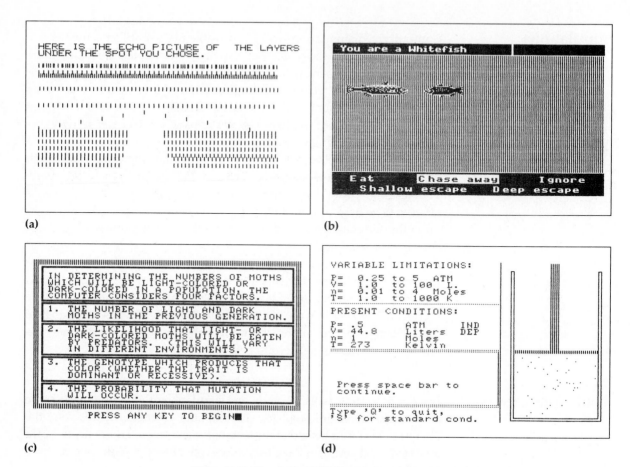

(a)

(b)

(c)

(d)

Figure 2.7. Simulations that let students perform scientific experiments. In screen (a), GEOLOGY SEARCH (Tom Snyder Productions/McGraw-Hill), the student has just performed a seismic blast. In screen (b), the student is waiting to see which fish eats which in ODELL LAKE (MECC). In screen (c), factors are presented that affect outcomes in MOTHS (Diversified Educational Enterprises). In screen (d), a chamber from IT'S A GAS shows the relationships between temperature, pressure, and volume (Diversified Educational Enterprises).

to teach science skills. Simply stated, the computer should be used in such a way that enhances the academic learning time of students in science classes. If an application enhances academic learning time, it will lead to improved science performance; otherwise, it would be better to use a different tool or strategy to promote learning. The computer can be integrated with instructional principles (especially mastery learning, direct instruction, and questioning strategies), to enhance academic learning time.

Finally, the computer can provide drills, tutorials, simulations, and tool applications in the science classroom. Subsequent chapters provide specific examples of these types of software.

REFERENCES

Aldridge, B. G. "Essential Changes in Secondary Science: Scope, Sequence, and Coordination." *NSTA Report* (January/February 1989): 1,4–5.

Anderson, C. W. "Incorporating Recent Research on Learning into the Process of Science Curriculum Development." A commissioned paper for the Biological Sciences Curriculum Study, 1987.

Andersen, H. O. "Minds-On Science." *Hoosier Science Teacher* 12 (December 1986): 55–61.

Ault, C. R. "Time for Elementary Science on Principle: Part I." *Hoosier Science Teacher* 11 (February 1986a): 71–74.

Ault, C. R. "Time for Elementary Science on Principle: Part II." *Hoosier Science Teacher* 12 (September 1986b): 6–11.

Barell, J. "You Ask the Wrong Questions." *Educational Leadership* 42 (May 1985): 18–23.

Berliner, D. "The Half-Full Glass: A Review of Research on Teaching." In P. Hosford (Ed.), *Using What We Know About Teaching*. Alexandria, Va.: Association for Supervision and Curriculum Development, 1984.

Caldwell, J. H., W. G. Huitt, and A. O. Graeber. "Time Spent in Learning: Implications from Research." *Elementary School Journal* 82 (1982): 471–480.

Dashiell, D. "Math/Science Education: Starting the Long Road Back." *Education USA* Special Issue (10 January 1983): 150.

Gagné, R. M. *The Conditions of Learning*. New York: Holt, Rinehart & Winston, 1985.

Guskey, T. R., and S. L. Gates. "Synthesis of Research on the Effects of Mastery Learning in Elementary and Secondary Schools." *Educational Leadership* 43 (May 1986): 73–80.

Lawson, A. E., M. R. Abraham, and J. W. Renner. *A Theory of Instruction: Using the Learning Cycle to Teach Science Concepts and Thinking Skills*. National Association for Research in Science Teaching, 1989.

Levine, D. U. *Improving Student Learning Through Mastery Learning Programs*. San Francisco: Jossey-Bass, 1987.

McIntosh W. J., and D. L. Zeidler. "Teachers' Conceptions of the Contemporary Goals of Science Education." *Journal of Research in Science Teaching* 25 (1988): 93–102.

National Assessment of Education Progress (NAEP). "Released Exercise Set." In *The Third Assessment of Science* (08-S-08). Denver: Education Commission of the States.

Rivers, R., and D. Luncsford. "The Design of a Computer Problem-Solving Test for Secondary Students." Paper presented at the sixty-second annual meeting of the National Association for Research in Science Teaching, San Francisco, 1989.

Rosenshine, B. V. "Synthesis of Research on Explicit Teaching." *Educational Leadership* 43 (April 1986): 60–69.

Schoon, K. "Misconceptions in the Earth Sciences: A Cross-Age Study." Paper presented at the sixty-second annual meeting of the National Association for Research in Science Teaching, San Francisco, 1989.

Slavin, R. E. "Mastery Learning Reconsidered." *Review of Educational Research* 57 (Summer 1987): 175–213.

Sund, R., and L. W. Trowbridge. *Teaching Science by Inquiry.* Columbus, Ohio: Merrill, 1967.

Tobin, K. "The Role of Wait Time in Higher Cognitive Level Learning." *Review of Educational Research* 57 (Spring 1987): 69–95.

Vockell, E. L., and E. Schwartz. *The Computer in the Classroom.* Watsonville, Calif.: Mitchell, 1988.

Wang, M., and H. Walberg (Eds.). *Adapting Instruction to Individual Differences.* Berkeley, Calif.: McCutchan, 1985.

C H A P T E R 3

THE COMPUTER IN THE ELEMENTARY SCHOOL SCIENCE CLASSROOM

TEACHING EPISODE 1

MARY JONES, A FIFTH-GRADE teacher at Hickory Creek Elementary School, is beginning the fifth in a series of physical science lessons with her students. Today, they are to learn the concept of density and how to measure it. She displays some packages of food, each showing the mass in grams. The packages are in a single row, lined up by the size of the package, smallest to largest. She asks her students if the volume of the packages is related to their masses. Most students are silent. One student responds by saying he doesn't know what mass is. "We studied that four weeks ago. You should know that!" Mrs. Jones counters. Another student ventures an answer: "I think that the larger the volume, the larger the mass." Mrs. Jones responds, "That's not a bad answer, but it's not exactly right."

Mrs. Jones then writes the term *density* on the board and asks the students if they know what it means. Susan says, "My mother sometimes says that my father is dense. Is that the same kind of word?" "No, that word means that someone has difficulty in understanding something," replies Mrs. Jones. "I want you to think of a word that has something to do with matter, like a dense forest or a dense population."

Mrs. Jones asks the students to turn to page 56 in their science books. The page pictures a double-pan balance. The right pan holds a nail, and the left pan holds a block of wood; the pans are balanced. She asks, "What do you observe from the picture?" Ramon responds, "The nail is heavier than the wood." "But notice that the scale is balanced; this means that they are both the same mass," says Mrs. Jones. Susan responds, "But don't nails weigh more than wood?" "It depends on the size of the nail and the size of the piece of wood," says Mrs. Jones, who then asks, "Do the nail and the block of wood have the same volume?" John says, "Yes." "Why do you say that?" asks Mrs. Jones. "Because the scale is balanced," answers John. "But that is because the mass of the nail and the wood is the same," responds Mrs. Jones.

She then asks the students to read the page silently. The text points out that the wood has a bigger volume than the nail, but that the nail would have more matter if it were equal in volume to the wood. Then the text defines *density* as the mass in a certain volume of matter and gives a verbal example.

After the students have read the page, Mrs. Jones asks, "Can anyone define the term *density*?" Amy responds by reading the definition in the text: "Density is the mass in a certain volume of matter." "Good," responds Mrs. Jones, "that's exactly right!" Mrs. Jones then holds up two identical cubes of material—one plastic foam, the other clay. She asks the students,

"Which of these cubes has the greater *density*?" Tyrone shouts, "The clay weighs more." "Yes, but does it have the greater density?" "Probably so," responds Tyrone. This ends the lesson for the day.

The next day, Mrs. Jones asks her students to turn to page 58, to read the page, and to look at the picture. On that page, the text asks the reader to compare the densities of different kinds of solid matter to the density of water. There is a picture of two kinds of soap in a large container of water. One bar of soap is floating at the top, and the other is resting on the bottom. The passage states that solid matter sinks in a liquid if the solid has a greater density than the liquid; otherwise, it floats. The text then tells the reader that the two kinds of soap have different densities and asks the question, "Which one has more matter in it than water?" After students have read the passage, Mrs. Jones asks them to answer the question. Ralph responds that "both soap bars have more matter than water because they are solid." Mrs. Jones states that "while most solids are more dense than liquids, some are not. The soap that sank to the bottom of the container was more dense than water, and that's why it sank. The soap that floated at the top was less dense than water."

"Now," says Mrs. Jones, "name some solids that are less dense than water." Amy responds, "Water is real light. I can't think of anything lighter than water." "Think about some things that float on water," says Mrs. Jones. "What about battleships?" John responds. "No, ships are denser than water," says Mrs. Jones. "Then why do they float?" asks John. Mrs. Jones gets flustered but responds by saying, "We are getting off the lesson. What I wanted you to say was ice or cork. Both of these are less dense than water." "But isn't ice just frozen water?" asks Susan. "Yes, but it gets less dense as it freezes." "That's weird!" comments Sammy.

At the end of the lesson, Mrs. Jones has all students copy the definitions for matter, length, mass, volume, and density from page 58. Most students do so. She then checks their definitions and asks the students to use the definitions to study for a test on Friday. The test is given as promised. It consists of the set of terms mentioned above with a mixed set of definitions. The students have to match the terms with their appropriate definitions.

TEACHING EPISODE 2

Mr. Smith, another fifth-grade teacher at Hickory Creek Elementary School, has almost the same experience teaching the section on density as did Mrs. Jones. At the end of the section, Mr. Smith tests his students. Many cannot match the definitions with their appropriate terms. He decides that a computer program that gives students practice in matching definitions with terms studied in the section would be helpful. He locates a drill-and-practice program on "Properties of Matter: Density" for elementary school children. One by one, throughout the day, he sends students who have experienced difficulty in defining terms to the computer for a short drill-and-practice session over terms associated with density. He then retests all students. The scores are more in line with his expectations.

TEACHING EPISODE 3

Mrs. Brown, a fifth-grade teacher at East Lansing Elementary School, has just begun a series of problem-solving experiments with her students, centered around the principle of buoyancy (adapted from McLeod and Hunter, 1987). She starts by displaying on tables sets of objects, some of which will float in water and some of which will not. She then asks her students to predict, on record sheets, which will float and which will sink. After students have filled out their record sheets with their predictions, she instructs them to drop each object into the container of water to see whether it sinks or floats. The objects include paper, several light woods, rocks of different sizes, plastic blocks of several densities, and several metals, including iron. Students then record their observations next to their predictions on the record sheets. In any case where they are inaccurate in their predictions, students try to determine why they were wrong.

There is a great deal of discussion among the students over one object. One plastic block has floated after most students predicted that it would sink. They are puzzled. Mary comments, "Somehow, this object is lighter than water. Maybe it has some air in it." Mrs. Brown says, "Interesting theory. How would you prove it?" Mary answers, "Make another block out of the same material without air in it." "Sounds good to me, Mary," replies Mrs. Brown.

"So why do you think some objects, like iron, sank and other objects, like balsa wood, floated?" asks Mrs. Brown. Leonard responds, "Because iron is heavier than balsa." Most of the class agrees with Leonard. Mrs. Brown then says, "Let's see if we can understand this property you call 'heaviness' or 'lightness' better. I want you to search our computer database on physical materials and look at the information on iron and balsa wood." Table 3.1 contains the result of the database search. The students have no trouble understanding all properties listed for each material, except for

Table 3.1. Database information about iron and balsa.*

Common name: Iron
Element, compound, or mixture: Element
Density: 7.86 (g/mL)
Color: Gray
Odor: Odorless
State at 25° C: Solid

Common name: Wood, balsa
Element, compound, or mixture: Mixture
Density: 00.1250 (g/mL)
Color: White yellowish brown
Odor: Odorless
State at 25° C: Solid

---

* From McLeod and Hunter, 1987.

their *density*. This is the concept they are being asked to develop. Mrs. Brown explains the concept to them by telling them that density is the amount of mass in a certain volume of matter. The *mass* is measured in grams (g); the *volume* of matter in milliliters (mL). Water then has a density of 1.0 because it has about 1 g of mass in 1 mL of volume. She then asks them to look at the densities of both iron and balsa wood. Two students note that iron's density is greater than 1.0 and it sank and that balsa wood's density is less than 1.0 and it floated. They then decide that materials with densities greater than 1.0 will sink in water and materials with densities less than 1.0 will float. Mrs. Brown asks them to create a list of solid objects from the database that have a density less than 1.0. At first the students go from screen to screen on the database, searching for the right materials. Then one student suggests that they ask the computer to print out all solid objects that have densities less than 1.0. Other students suggest that they might as well get a list of materials with densities greater than 1.0. Table 3.2 gives the results of the search.

Mrs. Brown now asks the students to first test their inference that materials with densities less than 1.0 will float by dropping into a beaker of water a piece of paper and some pieces of wood. The specimens do indeed float, as their densities predicted they would. She then asks the students to do the same with solid objects that have densities greater than 1.0. They do, and the evidence confirms their inference that such objects will sink.

Mrs. Brown then asks the students to go back to the computer and print out a list of solid objects ordered by density. After some effort, the students produce the list in Table 3.3. She then asks the students to make a prediction based on their inference. She asks whether a substance with a very low

Table 3.2. Sample list of solid objects with densities above and below 1.0, ordered alphabetically.*

Name	Class	Density	Color
Aluminum	Element	2.7000 (g/mL)	Silver
Chalk	Mixture	2.4000 (g/mL)	White
Iron	Element	7.8600 (g/mL)	Gray
Limestone	Mixture	2.8000 (g/mL)	Gray-variable
Paper	Mixture	0.9200 (g/mL)	Yellow
Wood, ash	Mixture	0.6833 (g/mL)	White
Wood, balsa	Mixture	0.1250 (g/mL)	White Yellowish brown
Wood, maple	Mixture	0.7310 (g/mL)	White
Wood, white oak	Mixture	0.7787 (g/mL)	White yellowish

* From McLeod and Hunter, 1987.

number like balsa wood will float lower in the water or higher in the water than something with a higher density number, like white oak. Sean confidently predicts that "balsa will float the highest because it is the lightest." James agrees with him, "Balsa has the lowest density number, so it will float the highest!"

Each group of students then tries several items from the list to see if their predictions are accurate. Most of them are pleased to see their predictions are accurate.

Table 3.3. Sample list of solid objects with densities above and below 1.0, ordered by density.*

Density	Name	Class	Color
0.1250 (g/mL)	Wood, balsa	Mixture	White Yellowish brown
0.6833 (g/mL)	Wood, ash	Mixture	White
0.7310 (g/mL)	Wood, maple	Mixture	White
0.7787 (g/mL)	Wood, white oak	Mixture	White yellowish
0.9200 (g/mL)	Paper	Mixture	Yellow
2.4000 (g/mL)	Chalk	Mixture	White
2.7000 (g/mL)	Aluminum	Element	Silver
2.8000 (g/mL)	Limestone	Mixture	Gray-variable
7.8600 (g/mL)	Iron	Element	Gray

* From McLeod and Hunter, 1987.

Mrs. Brown asks her students to predict which of the objects listed in Table 3.3 will float in alcohol (0.79 g/mL), in boiled linseed oil (0.942 g/mL), and in mercury (13.546 g/mL). She has them confirm their predictions in the laboratory with alcohol and linseed oil; but they only talk about what will happen in mercury, a toxic and expensive liquid.

Finally, Mrs. Brown asks her students why balloons filled with helium rise in the air and why they eventually sink. Jamie is quick with her answer, "Helium's density is lower than air, and so the balloon rises; and when the helium goes out of the balloon, it falls."

To test her students' understanding of the relationship between buoyancy and density, Mrs. Brown gives them a database printout of various gases and then has them answer the following hypothetical questions: Would a balloon filled with chlorine float in nitrogen? Would a balloon filled with hydrogen float in carbon dioxide? She is pleased with the results.

TEACHING EPISODE 4

Mr. Avery, a fifth-grade teacher at Engle Elementary School, uses the same physical science database that Mrs. Brown uses for a science lesson on density. He begins the lesson by lining up a set of objects on a table in front of the class. He says to the class, "All these objects have the same volume, but the one on the far left has the least mass, and the one on the far right has the most mass." Angela says, "What's mass?" The teacher responds, "The amount of matter in the object."

Mr. Avery continues, "Have any of you heard of the word *density*?" Julie says, "I have! Our book says that it's the amount of mass in a certain volume of matter." Mr. Avery responds, "That's correct. So starting with the object on the far left, each object has a greater density than the one to the left of it. I have a bowl of water. Notice that I have placed a label on each object, telling its density in grams per milliliter. OK now, Jacob, tell me if the balsa wood will float in this bowl of water." Jacob says, "I'm not sure." The rest of the class says it will float. Mr. Avery puts it in the bowl, and it floats. He says, "Notice that the density of the balsa wood is less than one. That means it will float. Andy asks, "But if it is less than one, shouldn't it sink?" "No, Andy," responds Mr. Avery, "if it is *less* than one, it should float." Mr. Avery then drops each object in the bowl of water so the children can see whether it sinks or floats and notes its density number. The children seem confused about what the number's relationship is to the object's ability to sink or float.

Mr. Avery then divides the students into small groups and asks each group to choose several objects to look up in a computer database. Each group is supposed to list ten objects and their densities and then tell whether each will sink or float in water. Unfortunately, many objects the students choose are not in the database; thus, most groups end up paging through the database, writing down the density numbers of the first ten or so materials. Because they don't understand what the number they have written down represents, they usually guess about whether it will sink or float based on the heaviness or lightness of the object.

After each group finishes its worksheet, Mr. Avery corrects the predictions that are not accurate and returns the papers. After all groups finish, he says, "Most of you had difficulty with this concept. Tomorrow you will write the definitions for mass, volume, and density, and then we will try to search the database again for new objects."

SCIENCE IN THE ELEMENTARY SCHOOL

These teaching episodes provide both bad news and the good news about science in the elementary school. Let's take the bad news first. Episode 1 is a typical example of how most elementary science textbooks and school environments encourage teachers to teach science to children.

1. The primary goal of the section on density became clear by the end of the teaching period. It was not for the student to develop an operational definition for density that could be applied in a variety of situations. If that had been the objective, the lesson would have been a failure because it presented the concept in a much too brief, verbal, and abstract manner for students who were mostly not developmentally ready to deal with it. The primary goal, revealed by the testing at the end of the chapter, was for the students to learn book definitions for density and other related terms.

2. While some lip service was paid to "scientific inquiry" in the questioning of students, the primary instructional model used by the teacher and the text was "information giver and authoritative source." Very little opportunity was given for students to manipulate materials to develop scientific problem-solving skills and accurate concepts of science phenomena.

3. The questions asked by the teacher indicate that she was, for the most part, following the book, looking for answers from the students that

were similar to those given in the teaching guide. When students did not respond appropriately, she was at a loss regarding how to proceed. At the end of the lesson, students, in general, used and maintained their previously held misconceptions or alternative conceptions about density. Before we become too critical of Mrs. Jones, we must remember that this is the curriculum she was given to teach and that curriculum supports the teaching/learning model with which she is probably most comfortable. She stayed close to the book because her own background in science is not particularly strong and she feels uncomfortable going beyond the explanations given in the text.

4. There are others who support Mrs. Jones in using this largely ineffective approach. The principal, who periodically evaluates Mrs. Jones's teaching, is comfortable with a teacher-as-information-giver approach. The principal feels that direct teaching of such concepts and skills is the most efficient because so many must be covered in each year. Mrs. Jones usually gets good evaluations when teaching science. The students seem cooperative and sometimes interested, especially during demonstrations. The custodian also appreciates the lack of cleanup needed with science taught from a book using pictures and teacher-demonstrated materials.

5. Since her district did not buy the kit of materials designed to be used with the text, Mrs. Jones had to "scrounge up" the packages of food for the beginning of the lesson and the cubes of plastic foam and clay for the end of the lesson. This was about all of the time she felt she had to give to the collection of materials. To organize any more complex student activity was out of the question.

6. Mrs. Jones and her principal both justify the use of their approach by stating that it can be used in the amount of time allocated to science in the curriculum. That time is limited to only twenty to thirty minutes a day. Only a text-based approach can be used in that limited framework.

That's the bad news. Now for the good news. Episode 3 is an example of a teacher and approach that is consistent with both the goals of good science teaching and with the intellectual capabilities of children. It deals directly with the children's misconceptions and builds on their natural curiosity.

1. The primary goal of the lesson on buoyancy also became clear by the end of this activity. Mrs. Brown's goal was for her students to apply the

concept (an operational definition of density) to a problem situation where the student used the concept to predict the buoyancy of one substance in another. The objectives associated with this goal require students to make accurate observations and inferences, make predictions based on relevant data, formulate hypotheses and design experiments to test hypotheses, and draw conclusions about the results of their experiments. Besides practicing these problem-solving skills, students develop an accurate concept of density and how it relates to the concept of buoyancy. Both the concepts and processes are central to science learning. The level (fifth grade) at which these skills were practiced and the concepts learned is realistic in terms of student intellectual capabilities. In addition, positive attitudes toward science as a human endeavor were generated, as well as curiosity and creative thought. Science attitudes such as feeling the need to verify data, willingness to have one's ideas questioned, and willingness to change an idea or concept when new evidence is presented are natural outcomes of such an activity.

2. More than lip service was paid to scientific inquiry in this teaching episode. The primary instructional model used by Mrs. Brown was one of teacher as facilitator. Mrs. Brown's guidance and questions led to real activities where students manipulated concrete materials that resulted in the collection of good data, valid inferences, and the confirmation of realistic hypotheses. A learning cycle approach was used, with students working in small cooperative groups. This approach consists of exploration (predicting and observing the sinking and floating of materials), term introduction (the discussion of the concept of density), and concept application (the use of the concept in predicting the buoyancy of new materials).

3. The questions asked by the teacher indicate that she was interested in having her students explore, predict, and infer events that they designed and controlled with her help. When students did not respond as she thought they would, she either used their alternative conception as a springboard to a new investigation or stimulated them to examine their idea from a different perspective. Mrs. Brown seemed comfortable with science as an active process, with students on center stage and with her in a role as a facilitator of learning. She was excited when students understood concepts and applied them accurately. She was even more excited when students developed good problem-solving skills and attitudes that supported them.

4. Mrs. Brown was provided with good background in science and in methods of science teaching. The curriculum in her school is activity-based science with an equal emphasis on science process skills and development of key science ideas and concepts. Very little emphasis is placed on the learning of factual knowledge in either teacher-made tests or standardized tests used by the district. The principal understands and supports what Mrs. Brown is doing. There is a science laboratory facility for the upper grades in East Lansing Elementary, with both equipment and supplies needed to support the program. A half-day aide is employed to help keep equipment and materials organized and replenished. The aide prepares the materials used by Mrs. Jones for the buoyancy activity. The only one slightly unhappy is the custodian, who had to clean up some linseed oil spilled on the floor.

USING THE COMPUTER IN ELEMENTARY SCIENCE TEACHING

None of the features of good elementary science teaching listed in the analysis of these episodes requires the use of the computer as an essential component. Indeed, much productive science teaching can be done using only concrete materials, science books, and a good teacher. The computer, as seen in Episode 3, *can* be used as an effective tool to enhance the teaching and learning of science. Note that in Episodes 2 and 4 the teachers used a computer, but there was no resulting improvement in learning. In most cases, when the computer improves learning, the benefit will arise (as Chapter 2 has shown) from enhanced academic learning time (ALT)—by providing an element that otherwise cannot easily be incorporated into instruction, by bringing the world into the classroom, or by using the speed and accuracy of the computer to allow students to spend more time on the learning task at hand.

Unfortunately, in Episode 2, the computer did not add appreciably to the benefit of the lesson. It did provide students some practice in the matching of terms. Most of this effort was wasted, however, because students did not really understand what they were learning and the computer program did not help them to do that.

In Episode 4, the computer actually detracted from the effectiveness of instruction. In that episode, the teacher simply used the computer to do Episode 1 badly. Instead of using the computer to focus student attention on a concrete activity, the additional activity probably subtracted from the intended ALT spent on the lesson. Students were engaged in a computerized activity, but this activity was not directed toward the lesson objectives.

In Episode 3, the computerized database added a great deal to the effectiveness of the lesson. It allowed students to quickly gather a great deal of data on a variety of materials. Many of these materials were either too dangerous, too expensive, or too difficult to work with in the allotted time. In minutes, students could retrieve and organize data using only one computer in the classroom that would in reality take hours or days using direct experimentation or noncomputerized reference sources. As a result, ALT was increased.

In addition, the integration of the computerized database into an activity-oriented science lesson encouraged students to practice scientific inquiry as they determined what questions to ask, what data answered them, and how they could best organize that data. When they asked the computer to produce various data sets, they acted like professional scientists organizing the results of their experiments to test a hypothesis.

The database was important to the objectives of the lesson in two ways. First, it allowed the students to greatly supplement the data collected from actual experimentation, so that patterns and trends in the data could be more easily identified. As a result, better inferences and hypotheses were formed. Second, students then used the inferences drawn from their manipulation of the database to design real experiments in the laboratory to confirm or not confirm those inferences. A *cycle* of learning was established where the student went from laboratory to database back to laboratory.

While it may initially be valuable to have a previously constructed database, most teachers will find it useful to have students add accurate information to existing databases or construct new databases of their own as they gain experience with their use. Teaching database management skills should usually be done outside the social studies class. Teaching them during class will be time consuming and may detract from other useful course activities. Database usage is an important higher-order thinking skill that can be generalized across many curriculum areas, and the science classroom may provide the ideal setting in which students can develop and practice this skill (Vockell and van Deusen, 1989).

In summary, it is important to note that the computer cannot save a poorly designed or poorly implemented elementary science curriculum. In some cases, it can even be used in a way that adds to the ineffectiveness of the curriculum. When integrated properly in a carefully designed curriculum, it can enhance the student's science ALT devoted to such areas as

(1) building a concept/knowledge base in science that can be used to accurately interpret real-world science experiences, (2) building a set of scientific problem-solving skills, and (3) developing information-processing and decision-making skills that can be transferred into many areas of the curriculum. Computer usage in elementary science has the potential to extend the effectiveness of the teacher in providing and organizing concrete problem solving; in offering data-gathering experiences for students; and in increasing skills, understandings, and positive attitudes related to the use of technology in one's world.

SPECIFIC COMPUTER APPLICATIONS IN ELEMENTARY SCIENCE

Harty and co-workers (1988) in a survey of the use of drill-and-practice, tutorial, simulation, and problem-solving software in elementary science found that students infrequently use computers to learn science at the upper elementary level and even more rarely in the lower grade levels. Drill and practice was used most frequently at both levels.

Three factors seem to be limiting the use of science software at the elementary level. First, there is a lack of software, especially high-quality software. *Electronic Learning* magazine recently interviewed Robert Haven, editor of The Educational Software Selector (TESS), concerning current trends in educational software. The question: Is there a subject area in which there isn't enough educational software available? Haven's answer was, in part, "I think the single most obvious lack in the past, though I think it is beginning to improve, is in elementary science software. There is very little, and the publishers haven't seen fit to provide enough emphasis in this area" (Gosewisch, 1989).

The second factor is lack of computers in classrooms at the elementary level, especially the lower grade levels. Many elementary schools have established computer laboratories for both literacy and learning, but most science applications (database, simulation, and microcomputer-based laboratories) are easily and effectively used in the science classroom or laboratory with only one or two computers.

The third and most important factor is that most teachers who teach elementary science are not science specialists. In fact, many went into teaching so that they could deal with people instead of facts. Unfortunately, many of them do not even like science. The good news is that almost all of them are susceptible to persuasion and can teach science effectively if they are given the materials and shown how to do it. Our experience has been that science is the curriculum area in which elementary teachers are least likely to seek out good software and use it correctly. To overcome this difficulty, school systems and teacher training institutions must give teachers opportunities for hands-on experience with good science software that will enable them to be successful in teaching concepts and developing skills that they themselves may previously not have mastered or enjoyed.

The remaining portion of this chapter surveys the applications of the computer to elementary science (grades K–5). Applications are presented in functional categories—that is, categories of use that describe the function the program will play in the curriculum. The four functions are instructional, revelatory, conjectural, and emancipatory (Schibeci, 1987).

Instructional Use

In this category, the computer is used in a direct instruction mode where learning tasks are highly specified and highly sequenced, with clearly defined prerequisites (Schibeci, 1987). Knowledge, comprehension, and application objectives are emphasized in software designed for this category (Bloom, 1956). Computer applications in this category include *tutorial* programs and *drill-and-practice* programs.

Tutorial Programs

Elementary science tutorials deal mostly with factual knowledge and concepts found in the typical elementary science text series. Many deal with nature study or health science rather than science. Some of these tutorials are no more than page turners, with students pushing the return key at the end of each colorfully illustrated screen. The well-designed programs provide clearly defined instructional objectives, require student interaction with both text and animated graphic material that illustrates the concept or process being studied, and give useful corrective feedback when students choose wrong answers. For example, the FIVE SENSES SERIES (Marshware) (Figure 3.1) provides good graphics and prompts students with an appropriate reading passage when a wrong answer is given. In addition, an on-line dictionary is provided for words that students may not understand. A life science series from the Center for Educational Experimentation, Development, and Evaluation (CEEDE) at the University of Iowa also emphasizes student/program interaction, providing error-contingent feedback that is both textual and graphic. Besides the tutorial lesson, a pretest and mastery test are provided along with a teacher management system that keeps track of student performance.

(a)

(b)

```
Which of these gives off light rays?
    1.    a camera
    2.    a flashlight
    3.    a tree
Enter your choice (1, 2, or 3) and
then press RETURN. 1

No.  Let's review.

........................................ Press RETURN.
```

(c)

(d)

Figure 3.1. Screen (a) shows a tutorial explanation of light rays from THE EYES HAVE IT of the FIVE SENSES SERIES. Screen (b) shows that program's on-line dictionary. Review questions, screen (c), and an appropriate reading passage, screen (d), are also provided.

There can be two main problems with science tutorials. First, the concepts taught by some of these programs are too abstract to be effectively learned by elementary school children. For example, a program on the molecular nature of matter claims that it is appropriate for grades 4–6 (Figure 3.2). It would be very difficult for fourth, fifth, or even sixth graders to generate meaningful understandings when presented with graphic illustrations of atoms and molecules. These programs are a reflection of the elementary science texts to which they refer. Both texts and computer

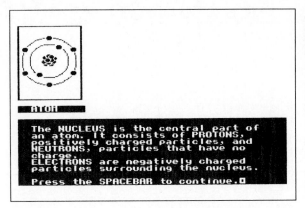

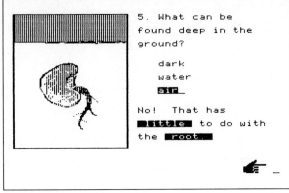

(a)   (b)

Figure 3.2. Screen (a) is from MATTER (Decision Development Corporation), showing a presentation that is too abstract for elementary students, and screen (b) is from GROWING SERIES: SEEDS that reinforces the misconception that air has little to do with root growth.

programs attempt to teach concepts that are not developmentally appropriate for the intended audience.

Second, teaching science concepts only in a direct, verbal fashion to young children requires simplification of the concept. When the concept is simplified, important features are left out, often resulting in the reinforcement of a misconception on the part of the child. GROWING SERIES: SEEDS (Profiles), for example, implies in its feedback to a test question that seeds in the ground only need water, not air, to germinate. A misconception that air is not found underground is likely to be reinforced in the student.

The skills and concepts that form the backbone of the elementary curriculum are usually best taught using either manipulation of concrete objects or some combination of concrete manipulation, simulation, data/information processing, and tutoring. Computer programs that combine simulation with computerized tutoring in an exploratory setting and then use the resulting insights in the formulation and application of a scientific rule can be helpful, especially when used in the application phase of the learning cycle. For example, the SCIENTIFIC REASONING SERIES (Figure 3.3) designed by Alfred Bork for IBM and compatible computers, based on the Elementary Science Study (ESS) unit entitled "Batteries and Bulbs," provides a good simulation of electrical circuits that allows students to discover patterns in the behavior of electricity (Bork et al., 1988).

Figure 3.3. Screen from SCIENTIFIC MODELS: BATTERIES AND BULBS, based on an ESS unit of the same name. Students discover patterns in the behavior of electricity.

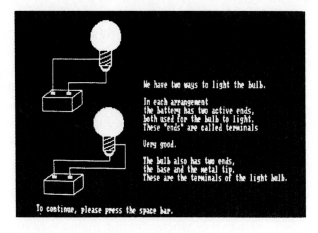

Drill-and-Practice Programs

Of the drill-and-practice science programs available for grades K–5, most are designed for the review of content presented in elementary science texts. Typical of these programs are MIND GAMES (Diversified Educational Enterprises), THE GAME SHOW (Advanced Ideas), and SECRETS OF SCIENCE ISLAND (Grolier). All have students answer questions in life, physical, or earth science areas to compete against others or against themselves in a game format. In MIND GAMES (Figure 3.4), students advance around a game board on the screen, by answering questions correctly. An editing feature allows educators to add new questions. In THE GAME SHOW, students are provided with clues to science words in a "Password" game format, receiving the maximum number of points for guessing the word on the first try. Science topics (words and clues) are

Figure 3.4. Game board from MIND GAMES.

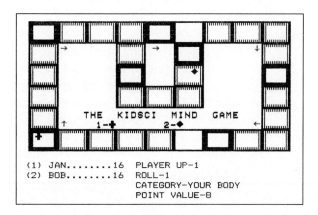

provided for each grade level (e.g., "Our Solar System" for grade 3). Teachers can add or modify topics. In the SECRETS OF SCIENCE ISLAND, students explore a "Troll Forest." They move through the forest by answering questions using a reference book provided with the software.

As noted for the tutorials, many of the concepts and terms practiced are not developmentally appropriate for primary grades (e.g., using "Our Solar System" as a topic for third grade) and, in some cases, actually support or teach misconceptions. These programs may have limited usefulness.

More useful at the early elementary level are programs that emphasize practice in the application of problem-solving skills fundamental to science in a game or puzzle format. These programs can be effective as supplements to activities where children manipulate real objects, either as additional practice or as a mechanism for evaluating student progress on such skills. Included in this category are programs such as THE POND (Sunburst), GERTRUDE'S PUZZLES and GERTRUDE'S SECRETS (The Learning Company), and MYSTERY OBJECTS (MECC). THE POND emphasizes pattern recognition, an important foundation skill in science problem solving. Both GERTRUDE programs and MYSTERY OBJECTS emphasize classification and ordering, important elementary science skills.   A preschool game called DINOSAURS (Advanced Ideas) encourages young children to sort and match dinosaurs on the basis of the dinosaurs' attributes.

Upper elementary programs (grades 3–5) that emphasize practice in higher-level science skills—such as making inferences and predictions, formulating hypotheses, testing hypotheses, and making generalizations—include programs like KING'S RULE, ANT FARM, and THE INCREDIBLE LABORATORY (Sunburst). Programs that teach such higher-order thinking skills and strategies for using them are discussed in Chapter 6.

Revelatory Use

In this category, the computer is used to mediate between the student and a qualitative or quantitative model of a real-world situation. These programs supplement concrete experiences or simulate concrete experiences when such experiences are too time consuming, dangerous, expensive, or remote from the classroom. Comprehension, application, analysis, and synthesis objectives can be taught with these programs. The computer applications in this category include *simulations*, *realistic games*, and *role playing*. Research indicates that the use of computer simulations in science has the potential to teach transferable problem-solving skills (Rivers and

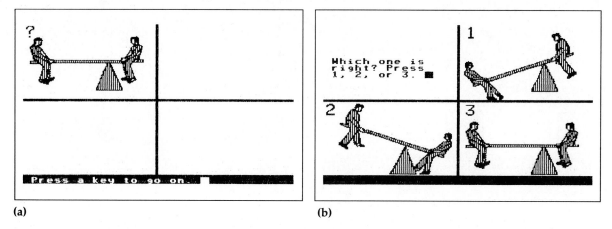

(a)   **(b)**

Figure 3.5 WHAT WILL HAPPEN? is designed to help primary students make predictions about physical and life science phenomena. Screen (a) presents a lever problem and asks the student what will happen next. Screen (b) presents choices that represent the possible predictions.

Vockell, 1987) and may be better than traditional instruction for teaching science concepts and understandings (Wise, 1989).

Simulations

Two good examples of simulations for elementary students are WHAT WILL HAPPEN? and ELECTRICAL CELLS (Macmillan). In WHAT WILL HAPPEN?, the student explores the phenomena of light, levers, and the effect of gravity on the motion of moving objects and on the growth of plants (Figure 3.5). The student is asked to observe and use prior experience to predict what will happen. In ELECTRICAL CELLS, the computer becomes a laboratory where students can discover how different metals can be combined in a battery to produce different amounts of energy (Figure 3.6). The "energy" can be used to "animate" battery-powered toys in the simulation. In both simulations, the student has an opportunity to develop both science process skills (observation, making inferences, formulating hypotheses, testing hypotheses, and making conclusions) and accurate concepts concerning light, electricity, and motion.

Realistic Games

ODELL LAKE (MECC) (Figure 3.7) is a realistic game that encourages students to determine predator/prey relationships between various fish and other organisms in an aquatic environment. Students develop skill in

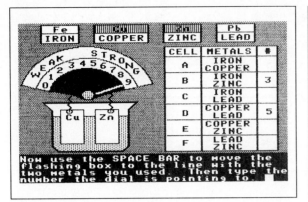

(a)

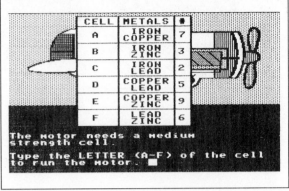

(b)

Figure 3.6 ELECTRICAL CELLS tutors students on the makeup of electrical cells, by allowing them to construct various types of cells and then use the cells to animate various toys. Screen (a) shows the student recording the strength of a particular cell. Screen (b) is directing the student to choose the appropriate cell to run the toy airplane.

making observations and inferences, as well as formulating a scientific model for a food chain or web in a lake. The game's motivating features, such as attractive graphics and competition between groups on point totals given for correct predictions, make this an exciting game for young students to play. THE BODY TRANSPARENT (DesignWare) offers a set of games that simulate the dissection of a human body (Figure 3.8). In the process of moving the various parts to their correct position in the human body outline, the student begins to understand their anatomy, position, and function.

Role Playing

Four examples of role-playing situations—programs that allow students to go beyond the normal classroom environment—are EXPLORING TIDEPOOLS, ANIMAL TRACKERS, PLANT DOCTOR, and WOOD CAR RALLY. In Walt Disney's EXPLORING TIDEPOOLS (Disney Education Enterprises), students explore a colorfully animated tidepool by moving objects and organisms and observing what happens as a result (Figure 3.9). A field guide is included in the program so that students can identify the organisms they find. Students also get a chance to create their own tidepools. In ANIMAL TRACKERS (Sunburst), students explore three different habitats and try to identify the animals that live in them by examining

ODELL LAKE: Lesson Based on a Descriptive Learning Cycle

This activity is based on ODELL LAKE, an upper elementary/middle school science simulation described in this chapter. The objectives of the lesson are that students will

- Use observations of the predator/prey relationships in a specific habitat to construct a food web of that habitat.
- Use the food web they constructed to predict the behavior of sets of organisms that are a part of the web.

Teacher's Guide

1. Introduce the problem to students by saying: "Your job will be to take the role of various fishes in Odell Lake so that you can determine the predator/prey relationships in the lake, that is, 'who eats who.' The predator is the one who eats, and the prey is the one who is eaten. After you identify all these relationships, you should be able to construct an Odell Lake food web and use it to predict what will happen when one organism meets another in the lake."

2. Give students Worksheet 1: Predator/Prey Relationships. In this sheet, students are required to fill in two things for each fish: first, what that fish likes to eat, and, second, what likes to eat that fish.

3. Have students begin to fill in their worksheets as you use the "playing to learn" section of the computer simulation as a whole-class demonstration to show how one fish responds to all other creatures near and around Odell Lake. Demonstrate enough interactions so that students will be able to complete their worksheets for that fish. Be sure to ask students questions like, "Do we have all of the information needed to fill in the chart completely for the Blueback Salmon?" "Do we know yet all the creatures that eat the Blueback?" "What do we need to know to finish our chart?"

4. Once you are sure that students are familiar with the program and the strategy they need to determine the predator/prey relationships, break up your class into heterogeneous groups. Have each group work on the computer simulation to fill in the chart completely for all other fish. Assign each student in each group a particular role. One should be an operator, one a recorder, one that keeps track of which fish have been done, and one that makes sure that all blanks are filled in for one fish before the group moves on to the next one. It might be wise to rotate the jobs once or twice during the session so that each student can play an active role. *Note:* This step can be done with large groups or the whole class if only one computer is available.

Continued

5. Have students share their completed Worksheet 1 in a whole-class discussion. Ask each group to report on a particular fish. Make sure the other groups check to see if the reporters have communicated accurate information. If not, go back to a demonstration of computer simulation to resolve the conflict, rather than give the answer yourself.

6. Give each person Worksheet 2: Odell Lake Food Web. Instruct each group to use Worksheet 1 to help them draw arrows from each fish to its favorite food or foods. Have one student suggest where to draw the arrow for the Blueback Salmon. Then have each group complete its food webs.

7. Place one group's food web on the board or on an overhead projector. Have other groups compare theirs with the demonstrated one. Resolve conflicts by going back to Worksheet 1. Make sure all groups have accurate webs.

8. Now have each group, one at a time, use its food web to play the portion of the computer simulation that is labeled "playing for points." In this mode, the computer places a time limit on each decision you must make as a fish (whether to eat the other fish, ignore it, escape, or dive shallow or deep). Points are awarded for accurate decisions. It would probably be wise to begin with a preliminary round. After the groups are functioning well, a serious round could begin. Friendly and lively competition will develop between groups.

Evaluation

As an evaluation activity, have students individually construct a food web based on information from another habitat that you have given them. Then have them predict what will happen in a series of encounters between organisms in that web.

graphic evidence of the presence of an animal (Figure 3.10). They record observations in a computerized notebook and then use an on-line field guide to identify their animals. In PLANT DOCTOR (Scholastic), students act as the plant doctor's assistant (Figure 3.11). Their job is to find out what factors in the plant's environment must be adjusted to make the plant well. They study each plant "patient," develop a hypothesis about why it is "sick," and test the hypothesis by performing controlled experiments. In WOOD CAR RALLY (MECC), students assume the role of scientists as they set out to determine the impact of the variables of car shape, car weight, lubrication, ramp length, and ramp angle on the distance a wood car will travel (Figure 3.12).

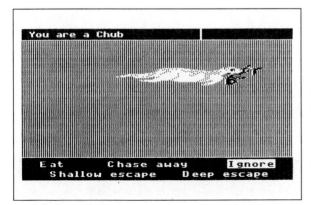

Figure 3.7 An otter (the predator) eats a chub (the prey) in a scene from ODELL LAKE. Students attempt to discover the predator/prey relationships in this lake by playing the roles of different fish that inhabit the waters of Odell lake.

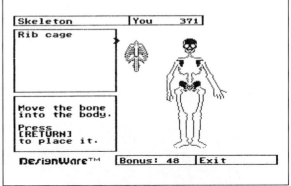

Figure 3.8 In THE BODY TRANSPARENT, the skeleton or organs can be assembled to show relationships among the bones and/or organs. Added material explains the location of bones and organs and their functions. This scene shows the placing of the rib cage into the skeleton.

Conjectural Use

In this category, computers are used to build interactive models of various science phenomena, so that hypotheses or configurations may be formulated and tested. The computer applications in this category are *microworlds* and *model-building situations*.

Microworlds

An excellent example of this kind of program is SIR ISAAC NEWTON'S GAMES (Sunburst) (Figure 3.13). This program is a microworld consisting of an object on a track. The track can be placed on earth (on ice, sand, or grass) or near the sun or out in space, allowing several different microworlds to be created and explored. The student must decide what to do to move the object around the track. In the process, the student develops a scientific model for motion in a particular microworld and uses that model to predict actions that will achieve the goal of movement around the track. This develops both science process skills and accurate concepts of real-world motion.

Model-Building Situations

The Weight and Density Group at the Educational Technology Center at Harvard University has developed a computer model-building program

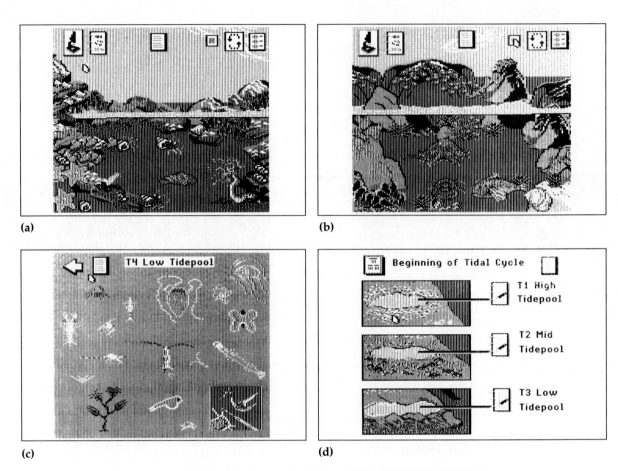

(a)   **(b)**

(c)   **(d)**

Figure 3.9 Screens (a–d) from EXPLORING TIDEPOOLS.

that allows students to "see" density. The program allows students to choose from several different kinds of materials of varied densities to build objects of different sizes. The screen displays as many as three objects, each depicted as a rectangle that is composed of rows of square units. Density is shown by the number of dots per square unit of size. The simulation works in conjunction with hands-on activities with objects of different weights, sizes, and densities. The teacher helps students move from the concrete experience to the computer simulation and back again, encouraging them to see the relation between the concrete objects they are manipulating and the more abstract computer representations. The experience of the Harvard

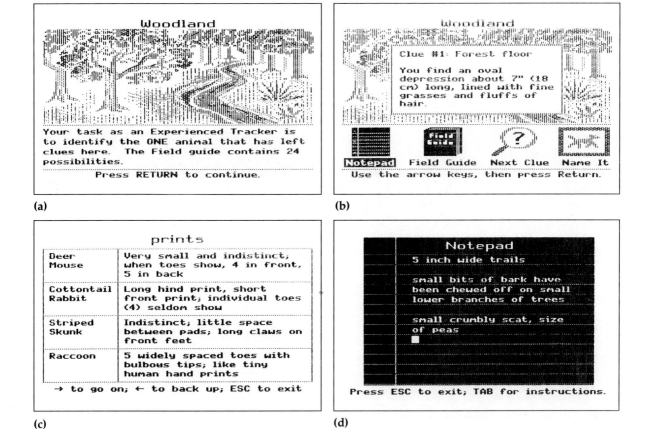

Figure 3.10 Screens (a–d) from ANIMAL TRACKERS.

group suggests that the "visualness" of the computer model aids students in forming mental models of the properties of matter.

Emancipatory Use

In this category, computers are used to free students from nonproductive work—that is, work that does not contribute to the lesson objectives. Examples of such work include repetitive calculation, technical setup, and organization of data into productive patterns. The computer applications here include *spreadsheets*, *word processing*, *microcomputer-based laboratories*, *database management*, and *telecommunications*. Spreadsheet programs can be used at the upper elementary school level to help students make repeated

(a)

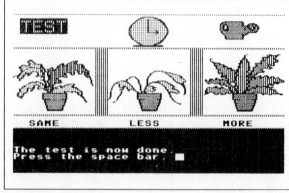

(b)

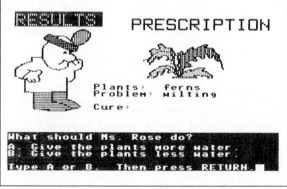

(c)

(d)

Figure 3.11 In PLANT DOCTOR, students act as the doctor's assistant, using the scientific method to diagnose and treat a particular plant's illness. First, the sick plant is presented to students, then they must try to guess what factor is contributing to its illness, screen (a). Next, they test that factor—in this case, water, screen (b). Then, the students judge whether this is really the factor and (if it is) whether the problem is too much or too little, screen (c). Finally, they help the doctor propose a prescription to make the plant healthy, screen (d).

arithmetic computations in tables of data they have collected from repeated trials in the same experiment or in different experiments. Word processing programs can facilitate writing of laboratory reports. Both spreadsheet and database printouts can be transported to the word processing program and patched into the student's laboratory report. These tools are discussed in greater detail in Chapter 7 of this book.

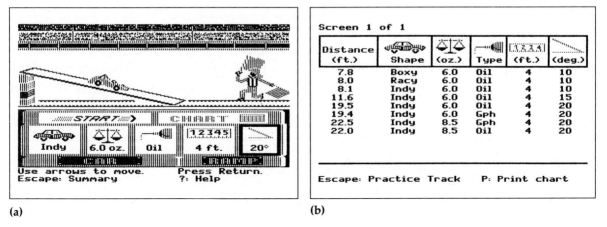

(a)   (b)

Figure 3.12 In WOOD CAR RALLY (MECC), the student sets up experiments (a) to test the characteristics of the model car and observes the results of several experiments on the data table (b).

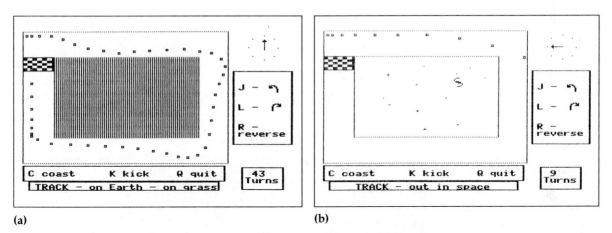

(a)   (b)

Figure 3.13 In screen (a), SIR ISAAC NEWTON'S GAMES offers a microworld in which an object can be placed on a track of grass on earth and the player kicks the ball around the track. In screen (b), the player is trying the same procedure in outer space.

Microcomputer-Based Laboratories

Microcomputer-based laboratories (MBLs) are discussed in detail in Chapter 8 of this book. PLAYING WITH SCIENCE: TEMPERATURE (Sunburst) (Figure 3.14) is an MBL designed for the elementary school and allows students to use up to three thermistors (temperature probes) to conduct over thirty experiments designed to fit into an activity-based elementary school curriculum. These experiments are simple enough for students at most elementary grades to handle. The program also encourages open-ended exploration. Data collection and storage are facilitated, and the accuracy of data collected increases with the use of this program. Students can immediately see the results when changing a variable in an experiment, many times altering the experiment as it is being conducted.

Database Management Programs

Several programs combine simulations or role-playing situations with databases so that students can use the database to solve the problem presented in the simulation. On-line field guides in ANIMAL TRACKERS and EXPLORING TIDEPOOLS, mentioned earlier, are examples of this kind of application. Others of this type are ANIMAL SCIENTIST (Scholastic) and GRIZZLY BEARS and WHALES (Advanced Ideas), two of the programs in the Audubon Wildlife Adventure Series. In these programs, on-line databases are used to help students, as they work through the simulation, identify animals.

A wide variety of stand-alone databases have been designed by the Bank Street College of Education to help students develop critical thinking and information-processing skills, as well as concepts and facts in science.

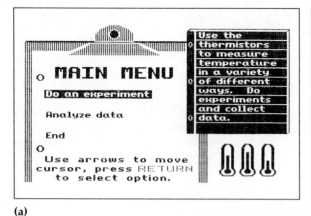

(a)

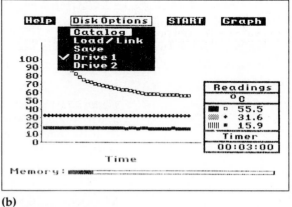

(b)

Figure 3.14 Screen (a) shows the menu for PLAYING WITH SCIENCE, and screen (b) shows part of a real-time graph with three thermistors, which are in three different beakers of water.

These databases require the use of a separate database management program, the BANK STREET SCHOOL FILER (all programs are distributed by Sunburst). They include ANIMAL LIFE (information on the physical characteristics, classification, behavior, food, habitat, reproduction, and adaptations of warm- and cold-blooded vertebrates), ENDANGERED SPECIES (all U.S. endangered mammals, extinct animals from 1600 to the present, and some of the most critically endangered species worldwide), SPACE (manned space missions, space probes, and a timeline file of important events in the history of space exploration), ASTRONOMY (file on the planets; weekly log of solar, lunar, tidal, and temperature data; and a timeline of important events in the history of astronomy), and CLIMATE AND WEATHER databases (weather maps and weather station). All are appropriate for use in the upper elementary grades. Extensive teacher's guides with suggested activities accompany each database.

Telecommunications

The term *telecommunications* refers to the process of communicating via modem between two or more computers in geographically different locations. Computer users are able to share programs or information in this way. The National Geographic Kids Network is a telecommunication-based science curriculum for grades 4–6. In this project, students participate in large-scale, cooperative experiments and share their results on the telecommunications network with other students throughout the world. Since the project began in 1986, five units have been developed by the Technical Education Research Center (TERC) and the National Geographic Society, funded by the National Science Foundation, the National Geographic Society, and Apple Computer. In one of the initial units, called ACID RAIN, elementary students use pH-indicator paper to collect rainfall acidity data from their geographic area. Each student's and each class's data are fed into a central processor, analyzed, and fed back to the participating elementary classrooms as maps and tables. Students then see how their own data fit into the overall pattern of acid rain, worldwide. A major outcome of such a project is the excitement generated among students (and teachers) as they become part of a scientific research "community" involved in contributing data to a worthwhile experiment.

The "Hello" unit of this package is designed to introduce the use of telecommunications in the classroom, to teach students how to collect and analyze data, and to provide a hands-on approach to learning the process of scientific investigation. ACID RAIN, WEATHER IN ACTION, and

WHAT'S IN OUR WATER?" are the investigative units presently available. "TOO MUCH TRASH?" will be available in the fall of 1991.

INTEGRATION OF SCIENCE WITH OTHER SUBJECT AREAS

Many of the programs discussed provide the opportunity to integrate science with reading, language arts, mathematics, and social studies. Indeed, many of the critical thinking skills emphasized in science programs can be applied to all these areas. In addition, when students write and read charts or laboratory reports, they practice and develop many important skills. The language used when students communicate with each other as they experiment should be clear, logical, and well organized. The opportunities to increase functional vocabulary when conducting science experiments is great. Science experiences provide an excellent "jumping off place" for language arts, reading, and mathematics lessons. The computer can enhance all these connections with the use of simulation, problem-solving, and database programs as well as telecommunications.

The VOYAGE OF THE MIMI (Sunburst), which is discussed in the next chapter, is also appropriate for some older elementary students. On a smaller scale, an example of a program that emphasizes such connections is THE DESERT (Collamore), which combines the simulation of a desert environment with both reading and writing activities. Using the program, students can identify desert animals and plants, explore habitats, gather information, and analyze desert food chains (Figure 3.15). Students can change and rearrange scenes on the screen, add text, and create new scenes, which can be saved. The computer does not allow students to create scenes that do not represent the habits of the creatures or the structure of the living desert.

Furthermore, most good reading and language arts teachers advocate a "whole language" or "reading and writing across the curriculum" approach. For this reason, many of the programs described by Whitaker and colleagues (1989) and King and Vockell (1991) support reading and language skills that contribute to the science curriculum. Similarly, many of the programs discussed by Dinkheller and colleagues (1989) support the integration of mathematics and science skills. Finally, Vockell and van Deusen (1989) describe numerous specific programs and strategies for transferring higher-order thinking skills across various curriculum areas.

Figure 3.15 THE DESERT (Collamore) provides a simulation of a desert environment in which students can identify desert animals, explore plants, and gather information about their relationships while practicing reading and writing skills. The screen shows a desert scene as the student has begun to construct it. The help menu is shown at the top of the scene.

SUMMARY

Computer applications have the potential to support an exciting, intellectually challenging, and relevant science curriculum for the elementary school. They also have the potential to enable teachers to be more efficient at providing poorly conceived, inappropriate, misguided, boring science education. For computers to significantly enhance science education, a more activity-oriented approach must be adopted, computer hardware and software must be provided for the classroom teacher, quality time must be committed to training the teacher in the teaching/learning models needed to implement such activity-based materials and software, and a support system of institutionalized resources must be provided. School districts in many areas have begun to take these steps. As they continue to do so, the potential of a strong and exciting curriculum in elementary science will begin to be realized.

REFERENCES

Bloom, B. S. (Ed.). *Taxonomy of Educational Objectives: Cognitive Domain*. New York: McKay, 1956.

Bork, A., A. Chiccariello, and S. Franklin. "Scientific Reasoning via the Computer." *The Science Teacher* 55 (May 1988): 79–81.

Dinkheller, A., J. Gaffney, and E. L. Vockell. *The Computer in the Mathematics Curriculum*. Watsonville, Calif.: Mitchell, 1989.

Gosewich, B. M. "Publishers Are 'Falling by the Wayside.'" *Electronic Learning* 8 (May 1989): 30.

Harty, H., P. Kloosterman, and J. Matkin. *Journal of Computers in Mathematics and Science Teaching* 7 (Summer 1988): 26–29.

King, R., and E. L. Vockell. *The Computer in the Language Arts Curriculum*. Watsonville, Calif.: Mitchell, 1991.

McLeod, R., and B. Hunter. "The Database in the Laboratory." *Science and Children*, 24(4) (1987): 28–30.

Rivers, R., and E. L. Vockell. "Computer Simulations to Stimulate Scientific Problem Solving." *Journal of Research in Science Teaching* 24 (1987): 403–415.

Schibeci, R. A. (1987). *Computers in Science Teaching: A Select Bibliography, 1983–86.* Project Description, Murdoch University, Australia. ERIC Document ED-287-688, 1987.

Vockell, E. L., and R. van Deusen. *The Computer and Higher-Order Thinking Skills*. Watsonville, Calif.: Mitchell, 1989.

Whitaker, B., E. Schwartz, and E. L. Vockell. *The Computer in the Reading Curriculum*. Watsonville, Calif.: Mitchell, 1989.

Wise, K. C. "The Effects of Using Computing Technologies in Science Instruction: A Synthesis of Classroom-Based Research." Paper presented at the annual meeting of the National Association for Research in Science Teaching, San Francisco, 1989.

CHAPTER 4

THE COMPUTER IN THE MIDDLE SCHOOL CLASSROOM

HOME ON THE RANGE

STUDENTS IN KENNETH FULLER'S science class at Faye Ross Junior High School in Artesia, California, were using the simulation BUFFALO to develop science problem-solving skills (Fuller, 1986). After a brief introduction to the activity, Mr. Fuller divided the class into groups, each group acting as a committee in charge of managing a buffalo herd for a national park. Each committee was given a range with a grazing capacity of 1 million buffalo and a starting herd size of 10,000, with the objective to breed the largest herd possible within a twenty-year period.

After experimenting for a while, most groups were able to increase their herds to around 50,000. Mr. Fuller noticed that almost all groups started with large numbers of calves and yearlings and equal numbers of males and females (not many farm families here). He then reported the following dialogue:

K.F.:   The groups with the largest herd started with more cows than bulls. Would one of you like to explain to the class why you did that?

STUDENT: You don't need a bull for every cow, and only cows have calves. The more cows you have, the more calves you get.

K.F.:   Very good. The next question is how many cows per bull do you need for the best results?

STUDENT: We can try different combinations and see which one works the best.

K.F.:   So let's design an experiment to determine the best ratio of females to males.

After an extended class discussion, Groups 1 to 9 ran the simulation with ratios of 1:1 to 9:1, respectively. The next day, the data were graphed and examined.

STUDENT: We still can't tell!

K.F.:   Why not?

STUDENT: Because the graph keeps going up. Maybe ten to one is better than nine to one.

K.F.:   What can we do?

STUDENT: Try some more ratios until the graph starts down.

K.F.:   OK. That sounds good. This time, Group one will work at ten to one and so on, to Group nine at eighteen to one. Then, we will add those data to our graph and see if it has peaked. Go to your machines.

On the next day, the graph peaked at 11:1, and Mr. Fuller congratulated the class. But the class wasn't ready to quit.

STUDENT: Mr. Fuller, even when we started with a high ratio, before many of our twenty years went by, the proportion returned to one to one. Wouldn't we get an even larger final herd if we kept the ratio the same the whole time?

K.F.:   That's an interesting question. How can we find out?

STUDENT: We can harvest enough males to keep the ratio the same.

K.F.:   Class, does that sound reasonable? It will require recalculating the ratio each year. And, because of the way data are presented in this program, you will have to take notes to calculate the number of yearlings and calves to be harvested from the previous year's crop.

After further explanations and discussion concerning procedures, the students returned to their computers. The complexity of the problem and the math skills necessary pushed many of the students to their limit; therefore, enough usable data could not be collected to answer the question. The results were inconclusive.

Then Mr. Fuller asked a very important question during the wrap-up: "What real-life factors have been ignored in making this simulation simple enough for us to use?" The students came up with such factors as bad weather, poor food and water supplies, and high incidence of predators. It is important for students using a simulation to understand the assumptions and limitations on which it is built.

MUMMIES, MAPS, AND MODELS

Students in the Sewickley (Pennsylvania) Academy Middle School have been part of a program to use the computer as a tool in various curriculum areas, including science (McQuade, 1986). The teachers of art, earth science, and computer science have developed an interesting collaborative project: the theory and practice of contour mapping. In both art and earth science, students use contour lines to make clay models. In art, students turn the human forms—represented in rough contour lines that define their shapes and look like head-to-foot bandages—into small clay sculptures. Students then slice their clay sculptures into cross sections and trace around each cross section on graph paper to make a set of contour lines. Students copy these lines onto a larger grid (increasing the scale) and trace them onto cardboard. The students cut out and assemble the cardboard cross sections, using scaled spacers to hold them apart. The result for every student is a sculpture of contours for which the original clay sculpture has served as a model.

In earth science class, students work in pairs on a scale model—about 1.5 x 1.8 m—of a volcanic island. The students decide on the topographical features, the scale, and the relationship of each segment of the model to the whole. They divide the island into squares 25 x 25 cm, and each pair of students is responsible for making and mapping the territory of one square in clay. The squares are then put together so that students can measure and record the elevations at regular intervals along the boundaries. This ensures that the numbers on the boundaries of adjacent squares agree when it is time to rejoin the clay squares into the island. The eventual goal is to create a composite contour map of the entire island, using a mapping program created for that purpose in the Logo programming language.

To measure and record the topography of the squares, a grid based on Cartesian coordinates has been created. This system of coordinates was chosen because Logo recognizes instructions containing referents on the x- and y-axes. Using straight pins and paper grids under their clay square segment of the island, students record the position and elevation of various points. They then use this data to make their contour maps.

So that they can use Logo to make their maps, students have learned to write Logo procedures with variables, to change their variables in a recursive program, and to set a limit on the recursion using a TEST command. They have also learned to use the system of coordinates and have had an opportunity to explore the use of the SETXY command.

The MAP procedure in Logo enables each pair of students to draw their contour lines on the computer by entering the coordinates at successive points along a line. They have learned to accommodate the distance between the locations of their elevation readings. To fit in the appropriate number of contour lines, they refer to their clay models as they work and check the numerical data with their visual interpretation of the terrain. Accurate coordinates, accurate typing, and regular storing of the work to date are the ingredients of success.

When the maps have been completed, they are printed, enlarged on a photocopier, and mounted on a bulletin board to form a composite contour map of the entire island. The separate clay squares of the island are reunited, the seams smoothed, and the finished model is put on display along with the map. Judging by the pride and the care with which the seventh graders have mounted their exhibition, the project has been a success. They have also scored well on a test of their knowledge of contour maps, the system of coordinates, and pertinent techniques of computer programming.

COMPUTERS IN MIDDLE SCHOOL SCIENCE TEACHING

In both of the classrooms described, the computer is used as an integrated tool to accomplish important educational objectives. In the BUFFALO scenario, qualitative thinking was enhanced by the use of the computer in a revelatory mode, allowing students to study a qualitative model of the population dynamics of a buffalo herd—something they would not be able to do otherwise.

This simulation accomplished several important objectives. First, with the teacher's assistance, students developed *scientific problem-solving skills* in a realistic setting (by developing hypotheses and designing experiments to test hypotheses). Second, students developed an accurate concept of population growth and of the major factors that influence it. This concept becomes a part of a *concept/knowledge* base that the student can use to accurately interpret real-world phenomena. Third, both the scientific skills and concepts that are learned while working on such problems make the student more adept in identifying and proposing worthwhile solutions to scientific problems that have societal (political, economic, and social) implications.

Situations similar to the BUFFALO simulation are abundant in the real world. For example, the world is faced with a severe decline in the African elephant population because hunters indiscriminately kill elephants for their ivory tusks. In a major South African national park, however, elephants are abundant because of planned management and selective hunting. In fact, the funding for this park is based on the sale of ivory from controlled hunting. If the sale of ivory were banned, this South African park would be in jeopardy because the ban would actually cause a decline in the elephant population. Students who understand the dynamics of such a problem will be in a much better position to help set public policy by proposing sound decisions based on good data, an understanding of population dynamics, and logical analysis of the variables involved.

In the second scenario, students used contour-mapping skills, usually associated with the earth sciences, to produce both a map and a piece of art. The computer served as tool to cut down on the labor-intensive portion of the mapping (the menial task of physically drawing the contour lines), allowing the students to focus on the *process* of accurately representing the topography of the island. In addition, students learned how Logo procedures could be developed and modified to produce desired results. In the future, students could use the skills developed in this lesson to design Logo procedures of their own.

The multidisciplinary approach used in this activity enables students to see relationships between concepts that have applications in different content areas. This approach may also facilitate transfer of problem-solving strategies from one area to another. Finally, the activity could be an effective introduction to the world of technology as applied to surveying and mapping. Interest in either art, computer programming, or surveying/ mapping careers could be generated with this activity. A multidisciplinary approach is especially useful in the middle school, where students are exploring various areas of study and are beginning to develop concepts, attitudes, and career choices related to them.

As these examples have shown, computer applications can be an effective tool to enliven middle school science programs and to

- Assist in the development of scientific problem-solving skills in a variety of science areas.
- Build a science concept/knowledge base that can be used to accurately interpret real-world phenomena.
- Develop skill in solving real-world science-related problems that have technological and societal implications.
- Develop positive attitudes toward science as a human endeavor and toward science-related careers.

The preceding examples and those discussed in the rest of this chapter show that the computer can make a significant contribution to the traditional middle school science curriculum. However, recent proposals for the development of a transitional educational philosophy for the middle grades require a rethinking of present organizational structures and curriculum scope and sequence. In general, these proposals would require that students be organized into communities within the school and that a group of teachers who specialize in various curriculum areas, including science, would be responsible for their learning. Teachers would form a team that would plan for the delivery of an integrated curriculum that emphasizes recurring themes, concepts, skills, and attitudes. Proposed reforms in secondary science scope and sequence would fit nicely into this framework (Aldridge, 1989). A set of core concepts in life science, physics, chemistry, and earth science would be defined and sequenced. All science areas would be addressed each year. These core concepts would be taught using a phenomenological approach, with students introduced to concepts via an inductive-style laboratory experience using the learning cycle described in earlier chapters. Connections with other subject areas such as language arts, mathematics, social studies, and art would be emphasized in the development of these activities.

The computer can be a valuable tool in the delivery of these laboratory experiences via simulation, microcomputer-based laboratories, and databases, as well as by providing tutoring and practice in the development and application of skills and concepts. Recurring themes, concepts, and skills can be enhanced by multidisciplinary packages that emphasize the integration of learning in more than one area, such as the VOYAGE OF THE MIMI series (Sunburst).

COMPUTER APPLICATIONS IN MIDDLE SCHOOL SCIENCE

Applications in this section are presented in the instructional, revelatory, conjectural, and emancipatory categories previously used in Chapter 3.

The remaining portion of this chapter surveys the applications of the computer to secondary science curriculum. Applications are presented in functional categories—that is, categories of use that describe the function the program will play in the curriculum. The four functions are instructional, revelatory, conjectural, and emancipatory (Schibeci, 1987).

Instructional Use

In this category, computers are used in a direct-instruction mode where learning tasks are highly specified and highly sequenced, with clearly defined prerequisites and objectives. The computer applications include *drill-and-practice* and *tutorial programs*.

Good middle school tutorials that have extensive branching and frequent student interaction are difficult to write and expensive to produce. The goal of most programs at this level is to develop understanding of more complex concepts such as cells, food webs, weather fronts, or energy. Such concepts require that tutorial programs follow the learning cycle by (1) introducing clear examples and nonexamples of the concept, (2) requiring the student to verbally state or recognize the concept, and (3) asking the student to apply the concept in a slightly more complex situation. Diagnostic feedback, especially during Steps 2 and 3, is especially important for accurate concept recognition and appropriate concept application.

Tutorials that recognize the need for such a learning cycle approach require a large written and visual database for Step 1, so that many examples could be given and patterns of attributes discerned. Tutorials require a great deal of flexibility and interactivity, including branching, in Steps 2 and 3, so that students can receive feedback at crucial points where they are attempting to make discriminations and to apply the concept.

Many tutorials have some elements of this approach, but few have been designed with all three steps in mind. SKY LAB (MECC) is one of them

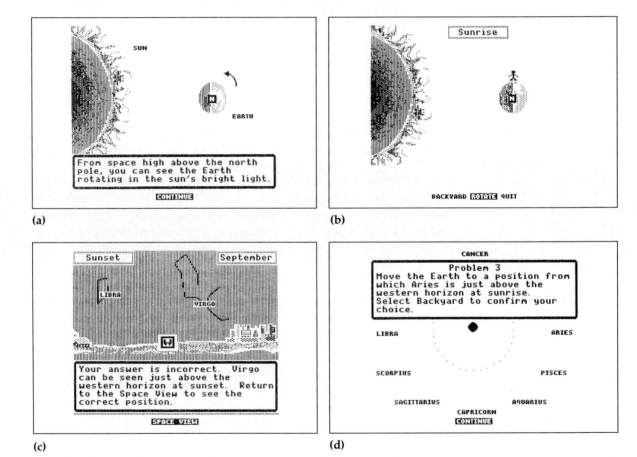

(a)

(b)

(c)

(d)

Figure 4.1. Screens from SKY LAB. Screen (a) illustrates the rotation of the earth. In screen (b), the student is exploring the motions of the earth. In screen (c), the student is applying the rotation concept. In screen (d), the student is applying the revolution concept. This sequence of activities follows the learning cycle model.

(Figure 4.1). SKY LAB calls itself an astronomy learning tool. It is nevertheless an interactive tutorial program modeled on the learning cycle approach. Each section of the program has three activities: "Demonstration Activity," "Discovery Activity," and "Test Your Understanding." The "Demonstration Activity" is a structured presentation of information about the concepts presented, such as the rotating earth. The "Discovery Activity" gives students an opportunity to use the options available in the demonstration without any prompting or guidance. The "Test Your Understanding" activity presents students with several problems to solve using the concept developed in previous activities. Although the presentation order of these activities suggests moving from structured presentation to free exploration, the teacher's manual suggests that the ordering of these activities could be under teacher and student control. These materials can easily be adapted to the learning cycle where the teacher briefly introduces a problem for the students to solve using the "Discovery Activity." The "Demonstration Activity" can be used to provide for term or concept introduction, and the "Test Your Understanding" activity can be used for concept application.

Most drill-and-practice programs at the middle school level are designed to introduce terminology to students receiving formal instruction in the life, physical, and earth sciences. These programs, for the most part, assume that if students can match verbal definitions with the appropriate term, they have some knowledge of the concept as well (Figure 4.2). These

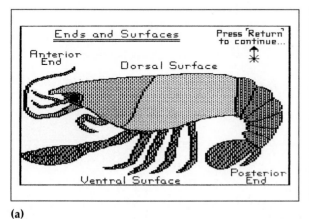

(a)

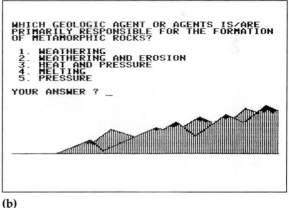

(b)

Figure 4.2 In screen (a), CRAYFISH DISSECTION (Cross Educational Software) shows the terms for the outer surface of a crayfish. Screen (b) shows a portion of the drill section from ROCK CYCLE (Ward's Natural Science).

programs function primarily at the knowledge and comprehension levels of Bloom's (1956) taxonomy of cognitive objectives. More important drills are those that help students apply concepts like mammal, metamorphic rock, and kinetic energy in new (to the student) settings. These programs often function at the application level of Bloom's taxonomy.

Many drill programs are games designed to motivate students to apply science concepts in interesting ways. WHO AM I? (Focus Media) enables students to identify plants and animals using specific text and picture clues in a "twenty questions" game format (Figure 4.3). The fewer clues needed to identify a plant or animal, the higher the score. Games are included for mammals, reptiles, insects, birds, trees, and flowers. Students who are studying animal and plant classification can apply concepts like mammal to a wide diversity of organisms. The concept becomes more broadly based, and critical attributes of mammals (hair, suckle their young) are differentiated from variable attributes (has claws or flippers), as the student progresses through the game. The teacher can edit the clues, making them easier or harder. This allows the program to be used with a wide variety of students.

Ventura publishes a series of identification games, some of which are appropriate for the middle school level. SENSES: THE PHYSIOLOGY OF HUMAN SENSE ORGANS (Ventura) is one of this series (Figure 4.4). This game provides motivating practice in identifying sense organ structures

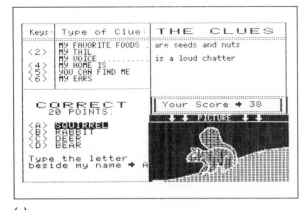

(a)

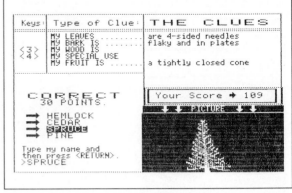

(b)

Figure 4.3 These two screens from WHO AM I? show the two levels of play in this game. In screen (a), the first level, the student looks at a picture of the organism and then identifies it. In screen (b), the second level, the student cannot see the picture unless all clues have been used.

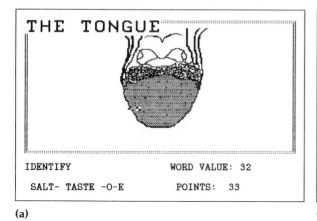

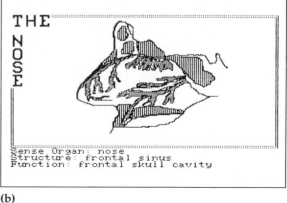

(a)   (b)

Figure 4.4 Screen (a) shows the identification game in SENSES: THE PHYSIOL-
OGY OF HUMAN SENSE ORGANS, a "Wheel of Fortune" game in which
students identify sense organs. Screen (b) shows the data retrieval file.

and functions that are represented in an anatomical diagram on the
computer screen. Each program includes a "Wheel of Fortune" game that
gives points for choosing a correct consonant or vowel in the name of the
structure or function. Bonus points are awarded for correctly guessing the
name of the structure or function. A "Data Retrieval Utility" is also
provided to give access to detailed information on the topics covered by the
game; a "Quiz Machine" provides an opportunity to practice associating
structures and functions in a multiple-choice test setting.

Revelatory Use

In this category, the computer mediates between the student and a qualitative or quantitative model of a real-world phenomena. These programs
can supplement laboratory experiences or take the place of laboratory
experiences when such experiences are too time-consuming, expensive,
dangerous, technical, or remote to be practical in the classroom. Computer
applications in this category include those that provide *simulations* and *role
playing*.

Simulations

Simulations at the middle school level help students learn and apply
higher-level science problem-solving skills. Students also develop accurate

conceptions of life, earth, and physical science phenomena that would otherwise be difficult to study in the school laboratory. Two life science simulations that accomplish these purposes are WEEDS TO TREES (MECC) and BOTANICAL GARDENS (Sunburst). The WEEDS TO TREES simulation presents the students with a plowed tract of land and nine kinds of plants that grow there (Figure 4.5). The students' task is to design investigations that will determine how each kind of plant grows and spreads. They can also design experiments to determine how each of the plant's eight variables influences how it interacts, succeeds, or dies in a given

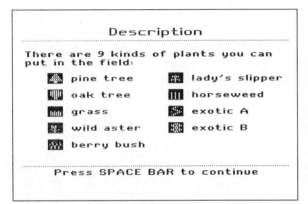

(a)

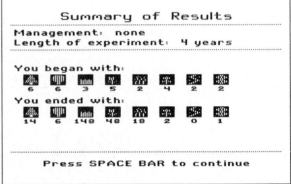

(b)

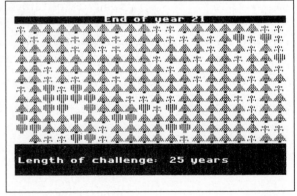

(c)

(d)

Figure 4.5 In WEEDS TO TREES, the student selects from nine species of plants (screen a) and then watches the plants grow, spread, and die out (screen b). In "Accept a Challenge" (screen c), the student is given a set of plants to keep alive for twenty-five years. Screen (d) shows the result of such a challenge.

situation. Students can also solve land-management problems using this simulation.

Two computer programs are provided. The first, "Explore the Ecology," allows students to explore relationships in an open-ended environment. The second, "Accept a Challenge," asks students to manage a tract of land so that a given set of plants will be maintained for a given number of years. Concepts addressed by this program include secondary plant succession, plant competition, and land management. Science problem-solving skills developed by the program include observing, predicting, inferring, designing experiments, and estimating.

BOTANICAL GARDENS provides a simulated greenhouse (Figure 4.6). Students can select seeds and grow plants under various conditions of light, heat, soil, and water. The results of each experiment are shown pictorially on the screen in the laboratory and graphically in a "graphing room." A genetics lab included in the program provides the students with the option of designing their own custom seeds that can then be grown in the laboratory. The process of designing a seed includes the selection of the plant's appearance and the generation of the hidden plant-growth parameters. Once a plant has been designed, it can be saved on the disk. Concepts addressed by this program include variables that affect plant growth: light, moisture, heat, and soil type. Science problem-solving skills developed by this program include controlling specified variables, looking for patterns in data, making inferences, and formulating and testing hypotheses.

Two good examples of physical science simulations for middle school students are WOOD CAR RALLY (MECC) and CHEM LAB (Simon & Schuster). WOOD CAR RALLY explores the physical science concepts of force and motion as the students use "wooden cars" to investigate how five variables—car weight, friction, shape, ramp angle, and length of ramp—influence the distance a car will travel once it leaves an inclined ramp. Three difficulty levels are presented, and students will be challenged by the two higher levels. CAR BUILDER (Weekly Reader Software) allows students to design, construct, refine, and test cars they build on the computer screen. They can design the chassis, engine, and suspension systems. The program includes a wind tunnel and a track to test the aerodynamics of the car as well as the racing and fuel capabilities. Although this program may be more useful in secondary industrial arts or applied mechanics classes, young students are often fascinated by the opportunity to see scientific principles at work in applied engineering settings that are relevant to their interests.

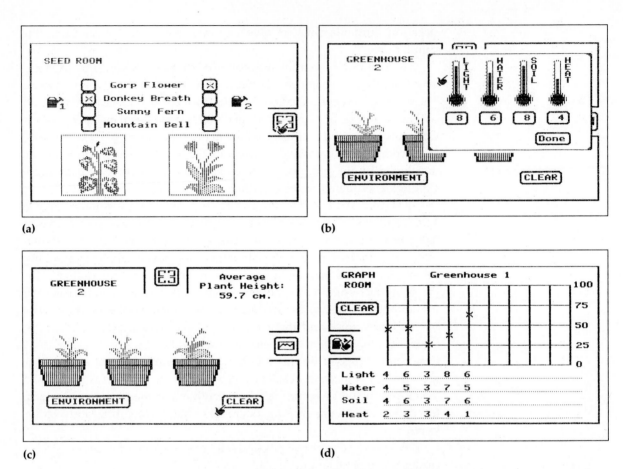

Figure 4.6 In BOTANICAL GARDENS, the student enters the simulated garden and selects the seed that will be grown (screen a). In the greenhouse, the student selects the growth environment (screen b). Screen (c) shows the results of one experiment on the plant called Donkey Breath; and screen (d) shows the graphical results of several experiments on the Gorp Flower plant.

CHEM LAB contains fifty different experiments that simulate actual chemical reactions that would be too dangerous or costly to do in a real lab. Students can simulate the making of such products as synthetic diamonds, gasoline additives, and rust removers. They use robot arms on the computer screen to manipulate lab equipment in special pressurized chambers.

THE EARTH AND MOON SIMULATOR (Focus Media) provides the students with laboratory-like control over a variety of realistic simulations with stop-action animation that show the relative paths of the earth and moon as they orbit the sun, the phases of the moon as seen from earth and from space, the rotation and revolution of the moon, sidereal and synodic months, and solar and lunar eclipses. Another animated sequence shows the cyclically changing tides as seen from space; another shows the shadow of the moon crossing the earth; while still another shows the phase of the moon as viewed from earth at the same time each evening. This program has the potential to help students not only develop concepts related to earth, moon, and sun relationships but also to develop skill in looking for patterns in data and making predictions based on those patterns.

Some useful middle school science simulations are not tied directly to specific science topics or content. The purpose of these simulations is to help students focus on the basic elements of experimental design. Using such simulations is much more useful than having students read the chapter in many science textbooks entitled "The Scientific Method." Students have considerable difficulty applying the generalizations found in such written passages. Usually, the generalizations are given verbally with few concrete examples. In most cases, there is no opportunity given for practice in applying the generalization.

A good example of this type of simulation is DISCOVERY LAB (MECC) (Figure 4.7). This program is designed to introduce students to the scientific processes of observation, experimental design, and hypothesis testing.

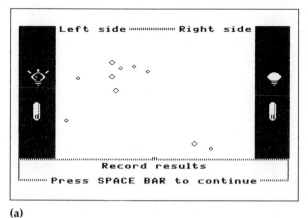

(a)

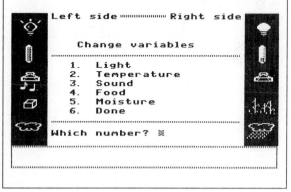

(b)

Figure 4.7 Screen (a) shows the process of testing an organism for its light preference in the training lab of DISCOVERY LAB. Screen (b) shows the list of environmental variables for the explorer lab and the challenge lab.

Students are challenged to design experiments to determine the characteristics of imaginary mystery organisms. The package contains three separate labs of varying difficulty. As students progress through the labs, they must control additional variables and analyze increasingly complex organisms. Each lab contains five types of organisms and a graphical chamber where the students can design and run their experiments. Students study the characteristics of an organism by changing the environmental conditions (light, temperature, sound, moisture, and food) in a chamber. Students determine the characteristics of the organism by observing its behavior under different conditions. When students are satisfied that they have discovered the behavior patterns of the organism, they compare their results with the actual characteristics of the organism. Possible redesigns of experiments are suggested when the students' findings do not match their organisms' true characteristics.

Role Playing

Three role-playing situations, which are good examples of programs that allow students to go beyond the normal classroom environment, are BACKYARD BIRDS (MECC), MINER'S CAVE (MECC), and GRIZZLY BEARS (Advanced Ideas) developed by the National Audubon Society. In BACKYARD BIRDS, students go on a bird-watching field trip (Figure 4.8).

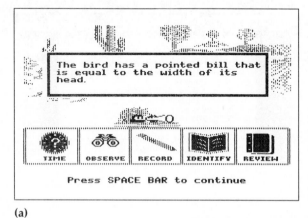

(a)

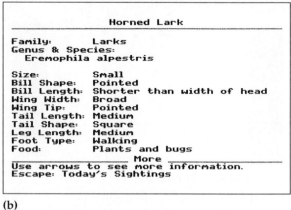

(b)

Figure 4.8 In BACKYARD BIRDS, students go on a simulated bird-watching field trip and observe bird characteristics (screen a). The on-screen field guide in screen (b) provides the characteristics of the various birds.

In "Field Trip," students are given the task of identifying a mystery bird. To accomplish the task, they must collect data about specific characteristics of the bird; they then organize and analyze the collected data by comparing them to the characteristics of a list of possible birds. Three factors affect the level of difficulty: the number of birds with which the students work, the number of traits the mystery bird has in common with the distractor birds, and the number of observations that students have noted before the mystery bird flies away. Using the field guide, students can research and review the program's database of over 120 North American birds. Besides learning to identify birds according to their physical characteristics, behaviors, and habitats, students practice the science skills of observing, comparing, relating, and inferring.

In MINER'S CAVE, students assume the role of miners to lift cartloads of jewels from the floor of a hidden cave to an elevator that will take the jewels to the surface (Figure 4.9). The original miners left four simple machines—the ramp (inclined plane), lever, the wheel and axle, and the pulley—that student miners can use to lift the treasure. The students' task is to determine which machine will work in the space available and to try to make maximum use of the amount of force available for each lift. From their explorations, students develop the ability to identify the variables that affect the efficiency and appropriateness of each machine. Science process

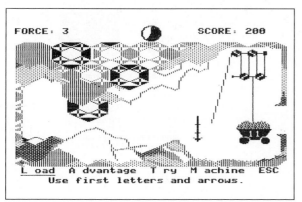
(a)

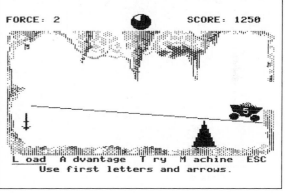
(b)

Figure 4.9 In MINER'S CAVE, students use machines to lift cartloads of jewels from the floor of the cave. In screen (a), a pulley setup lifts a load of 11. In screen (b), a lever lifts a load of 5. The student can set placement of the fulcrum, the number of pulleys, and the amount of the load.

skills developed in the program include observing, estimating, comparing, and inferring.

GRIZZLY BEARS (Figure 4.10) is a unique program that contains four interactive, graphically illustrated stories in which students become, in turn, researchers, rangers, detectives, and resource developers. In each story, students are confronted with decisions that will help them be successful in balancing the delicate harmony of the grizzly's wild habitat

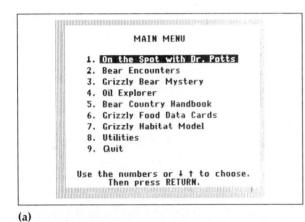

(a)

(b)

(c)

(d)

Figure 4.10 In GRIZZLY BEARS, students are researchers, rangers, detectives, and resource developers. Screen (a) shows the main menu. Screen (b) shows the "Bear Country Handbook," which lists the physical characteristics of the grizzly. Screen (c) shows the "Bear Encounters" module; screen (d) shows a "Grizzly Food Data Card."

within the civilized world of humans. The program makes three tools available to the students as they proceed through the stories. The first, "Bear Country Handbook," is a database on grizzly bear behavior, including physical descriptions, migration patterns, study techniques, history, and distribution patterns. The second, "Grizzly Food Data Cards," is a database of information on the foods that grizzlies eat, where these foods can be found, what they look like, and at which time of year they are most likely to be eaten. The third, "Grizzly Habitat Model," allows students to explore and discover the effects of timber harvesting, developing campgrounds and resorts, and oil drilling on grizzly habitat. This program encourages the student to develop scientific research techniques, practice decision-making skills, and develop concepts related to the grizzly, such as animal behavior, habitat, nutrition, resource development, and preservation of wild animals.

Conjectural Use

In this category, computers are used to build interactive models of various science phenomena, so that hypotheses or configurations can be formulated and tested. Computer applications in this category include *microworlds* (or problem-solving environments) and *model building*.

Microworlds

SIR ISAAC NEWTON'S GAMES (Sunburst), mentioned in Chapter 3, is also useful for older students. In addition, DISCOVERY: EXPERIENCES WITH SCIENTIFIC REASONING (Milliken) (Figure 4.11) consists of ten microworlds, or problem environments, and a tutorial problem intended to familiarize students with program procedures. Each microworld consists of a laboratory with various rooms, or screens, including tutorial rooms, practice rooms, challenger rooms, and "blackboards." In tutorial rooms, students are given an overview of the problem to be solved and any special aspects of the problem. Actual problem solving occurs in the practice room. The blackboard maintains records of problem-solving attempts conducted in the practice room. The students go to the challenger room when they are ready to solve the problem. The challenger room scores the solution and keeps track of the number of problems solved correctly. The students must solve three problems in a row to pass the challenge.

The first problem is "Open the Lock," in which students must analyze their own thinking in order to open the lock on the front of a safe. Other problem microworlds where students are required to develop scientific

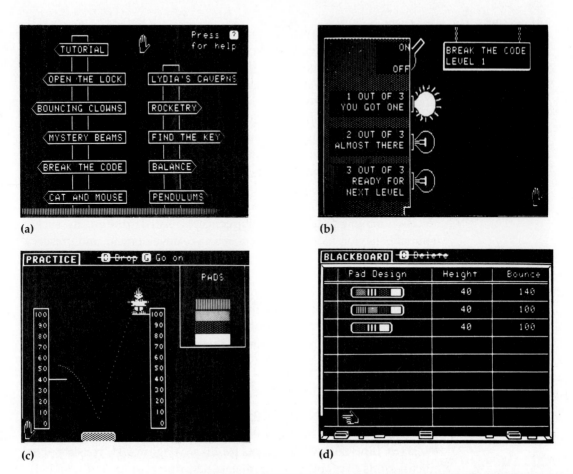

Figure 4.11 In DISCOVERY: EXPERIENCES WITH SCIENTIFIC REASONING, the signpost room (screen a) directs students to all microworlds. The challenger room (screen b) presents problems and counts the number of correct solutions. A practice room (screen c) is included with each problem environment. The blackboard (screen d) maintains records of problem-solving attempts in the practice room.

strategies include "Lydia's Caverns," "Bouncing Clowns," "Rocketry," "Mystery Beams," "Bird Balance," "Cat and Mouse," and "Pendulums." For the teacher, there is a DISCOVERY manager on a separate diskette, where the management program and student data are maintained.

Model Building

LAWS OF MOTION (EME) is a good example of a program in the model-building category. It allows students to build a scientific model for Newtonian motion while they experiment with the movement of a block down an inclined plane. Students can change the mass of the block and the bucket used to pull the block, as well as the angle of the plane and the friction and net force produced. The computer displays the results of the students' experiments in slow and real-time motion, stop-action "snapshots" at equal time intervals, and graphs and tables of the data. This program can easily be adapted for use in both the exploration and application phases of a learning cycle covering the concept of motion.

Emancipatory Use

In this category, computers free students from nonproductive work—that is, work that does not contribute to the lesson objectives. Examples of such work are repetitive calculation, technical setup, and organization of data into productive patterns. The computer applications here include *spreadsheet* and *word processing*, *microcomputer-based laboratories* (MBLs), *database management*, and *telecommunications*. Spreadsheet programs can be used at the middle school level to help students make repeated mathematical computations in tables of collected data from repeated trials in experiments. Word processing programs facilitate writing of laboratory reports. Both spreadsheet and database printouts can be transported to the word processing program and patched into the student's laboratory report. These tools are discussed in greater detail in Vockell and Schwartz (1988).

Microcomputer-Based Laboratories

The MBL, or "probeware" as it is popularly termed, is the subject of Chapter 8 of this book. An MBL can and probably should be an important feature of middle school science laboratory experiences. Such probeware gives students immediate feedback that is crucial to develop the ability to design useful experiments and to correctly analyze the results of experiments.

SCIENCE TOOLKIT: MASTER MODULE (school edition) (Broderbund) (Figure 4.12) includes light and temperature probes and program software that allow students to turn the computer into a timer, thermometer, light meter, or strip chart recorder. Students cannot print the strip chart, but they can scroll the chart back and forth on the computer screen. They can save data to their disks or print it, so that the data points can be

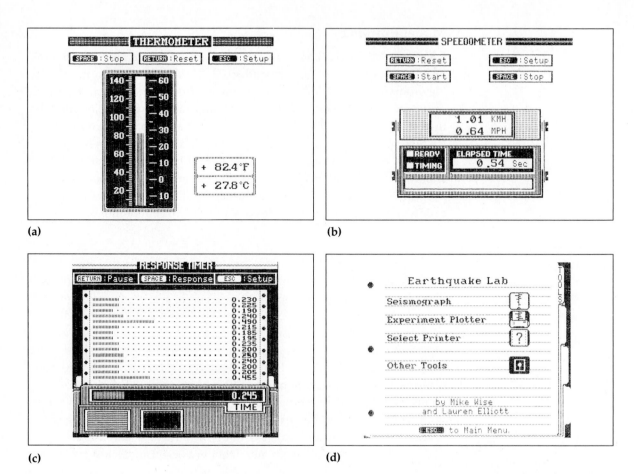

Figure 4.12 Screen (a) from SCIENCE TOOLKIT shows the rise in temperature of a cup of water on the thermometer. Screen (b) shows the speedometer that is used to time the balloon-powered car. Screen (c) shows the response timer from "Body Lab" module cell; screen (d) displays the "Earthquake Lab" menu.

used to create a graph similar to the one on the strip chart. A student manual outlines possible projects for students to complete using SCIENCE TOOLKIT; a teacher's guide is also included. Three enhancement modules are available: "Module I: Speed and Motion," "Module II: Earthquake Lab," and "Module III: Body Lab." Each adds capabilities to SCIENCE TOOLKIT (such as the printing of specialized graphs) and requires the use of the master module.

Two Sunburst probeware kits are also designed for use in middle school science. Students can discover the relationship between movement and lines on a position versus time graph in PLAYING WITH SCIENCE: MOTION. This probeware kit uses a motion detector to record movement that occurs in its path. This kit provides students with a powerful tool to help them analyze and study moving objects. EXPLORING SCIENCE: TEMPERATURE allows students to monitor temperatures ranging from –20° C to 120° C in up to 1-molar solutions. Built-in statistical functions help students calculate and graph data.

TEMPERATURE EXPERIMENTS (Hartley) allows students to simultaneously use two temperature probes to collect data. The computer shows temperature readings on a thermometer or plots a graph of the readings over a selected period of time. The graphs can be saved to disk and sent to a printer. An additional feature superimposes a graph, on the computer screen, from a single probe over another done at a different time, as long as both cover the same amount of time. The program also includes a tutorial with drill and practice on how to read thermometers, which is useful in developing this skill to a high degree of competence.

The experiments suggested in these materials are, for the most part, appropriate for middle school students. Open-ended exploration is encouraged. Data collection and storage are facilitated. Students can immediately see what happens when they change a variable in an experiment. They can also time the experiment or alter the experiment as it is being conducted.

Database Management

As students use a database manager to store and organize scientific information, they can generate and answer questions, formulate and test hypotheses, and critically evaluate the results of their inquiries. Along with the database manager and the sample data in the files that accompany it, solid teaching strategies and student materials are provided in a teacher's manual or guide. These aids are important because they greatly enhance the potential for productive use.

GEOWORLD (Tom Snyder Productions) (Figure 4.13) combines the simulation of a geological expedition with a database of mineral resources and a computer graphics world map accurate to 1 degree of longitude. Students use the data and map derived from the database to solve problems presented in the simulation. Topics addressed in the program include natural resource and mineral distribution, world geography, rock formation, and interpretation of maps, meters, charts, and cross-sectional diagrams.

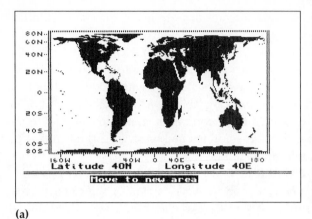

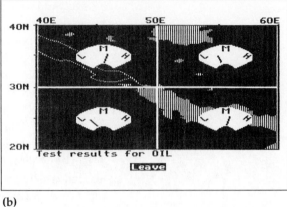

(a)   (b)

Figure 4.13 Screen (a) shows the world map used in the GEOWORLD simulation; screen (b) shows a geochemical test for oil in the Persian Gulf region.

A wide variety of stand-alone databases have been designed by the Bank Street College of Education to help students develop critical thinking and information-processing skills, as well as concepts and facts in science. These databases require the use of the BANK STREET FILER database manager. They include "Animal Life" (warm- and cold-blooded vertebrates), "Endangered Species," "Space" (space missions and a timeline file of important events in the history of space exploration), "Astronomy," "Climate," and "Weather" databases. All are appropriate for use in the middle school grades. An extensive teacher's guide with suggested activities is provided with each of the database files.

DATAQUEST: NORTH AMERICAN MAMMALS (MECC) is a database file filled with facts about the characteristics and behavior of ninety-eight mammals indigenous to North America. Students can easily use this database file to recognize taxonomic patterns and observe interrelationships among mammals. "Dataquest Composer" must be used to edit or add information to this file.

Telecommunications

The term *telecommunications* refers to a broad range of activities, including the linking via modem of geographically distant computers. The TERC Star Schools project combines a national telecommunications network, microcomputer, and video programming as vehicles for student-originated investigations in science and mathematics. The network enables secondary students to engage in collaborative projects, sharing data and results with other students, teachers, and scientists around the world. The testing of several newly developed curriculum modules is under way with students in grades 7–12. The project is funded by the U.S. Department of Education, Star Schools Program. It is a joint effort of the Technical Education Research Center in Cambridge, Massachusetts, and eleven outstanding educational institutions that serve as resource centers for participating schools. The centers provide teacher training and ongoing support to the participating schools, all linked through the telecommunications network. Network moderators and scientists offer feedback and guidance to teachers and students as they pursue challenging projects and explore new ways of teaching and learning.

All secondary classes begin with "Intronet," which introduces the telecommunications network and the concept of collaborative investigations. Classes then select one or more modules offered on the network and participate in a four-week core investigation. Modules designed to provide research skills include "Polling" and "Design." Science modules include "Radon," "Weather," "Trees," and "Tides." A mathematics module called "Patterns from Iteration" is also available.

All middle schools and high schools committed to improving their math and science programs are eligible to apply for participation in the TERC Star Schools Project. Schools from districts eligible for Chapter 1 funding for disadvantaged students are especially encouraged to apply.

INTEGRATION OF SCIENCE WITH OTHER SUBJECT AREAS

Many of the programs discussed in this chapter provide the opportunity to integrate science with reading, language arts, mathematics, and social studies. Critical thinking skills emphasized in science software can be applied to other contents areas. In addition, when students write their own reports and read other students' laboratory reports, many important reading and writing skills are developed and practiced. The language used when students talk with each other as they experiment helps them practice clear, logical, and well-organized communication, contributing to important language arts learning. The opportunities to increase functional vocabulary when conducting science experiments is great. Science experiences provide an excellent "jumping off place" for language arts, reading, and mathematics lessons. The computer enhances all of these connections with the use of simulation, problem solving, database programs, and telecommunications.

A specific software package that emphasizes such connections in a multimedia setting is the VOYAGE OF THE MIMI (Sunburst), which combines video, print, and computer modules into an educational adventure that weaves science, mathematics, and social studies into a fabric of learning (Figure 4.14). Thirteen dramatic video episodes tell the story of a

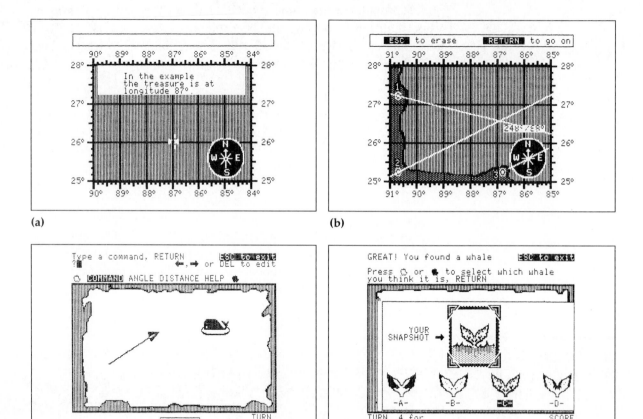

(a)

(b)

(c)

(d)

Figure 4.14 Screen (a) from the VOYAGE OF THE MIMI shows the explanation of longitude in the maps and navigation exercises. Screen (b) demonstrates learning to use a direction finder to locate one's own ship's position. Screen (c) is an example of using turtle graphics to move a boat to help rescue a whale. Screen (d) gives students practice in identifying whales from pictures of their fins.

research expedition aboard the seventy-two-foot ketch *Mimi* to study whales off the coast of New England. Along with the crew, students learn about whales, navigations, maps, and computers through their sea adventures. The student book provides full-color art and photographs and includes student activities that are based on themes covered in the episodes and expeditions and a glossary. An overview guide contains lessons and activities for each of the video segments and two large wall charts: a navigation chart of the Gulf of Maine and a poster of marine mammals. Four computer learning modules contain software and print-support materials, with themes derived from the *Mimi* adventure. They include "Maps and Navigation," "Ecosystems," "Introduction to Computing," and a probeware package called "Whales and Their Environment."

The SECOND VOYAGE OF THE MIMI (Sunburst) was introduced recently by Bank Street College of Education. This time, the *Mimi* and its crew lead students on an adventure to Mexico to study the ancient Maya civilization. This package includes twelve video episodes, overview guide, student book, and two computer modules: "Maya Math" and "Sun Lab."

Furthermore, most good reading and language arts teachers advocate a "whole language" or "reading and writing across the curriculum" approach. For this reason, Whitaker and colleagues (1989) and King and Vockell (1991) describe many programs supporting reading and language skills that contribute to the science curriculum. Similarly, Dinkheller and colleagues (1989) discuss many programs that support the integration of mathematics and science skills. Finally, Vockell and van Deusen (1989) describe numerous specific programs and strategies for transferring higher-order thinking skills across various curriculum areas.

SUMMARY

There is a variety of good software in science at the middle school level. Much of this software emphasizes the interconnections between concepts and skills within science and between science and other areas such as language arts, social studies, and mathematics. It is critical that such software be placed in a curriculum that emphasizes such relationships. It is also important to train teachers to make effective use of software and to provide an organizational structure in the school that will facilitate a team approach to teaching and learning. Telecommunications within schools and between school districts and other educational and scientific communities is essential in breaking down the barriers that keep us apart.

REFERENCES

Aldridge, B. G. "Essential Changes in Secondary Science: Scope, Sequence, and Coordination." *NSTA Report* 1 (January/February 1989): 4–5.

Bloom, B. S. (Ed.). *Taxonomy of Educational Objectives: Cognitive Domain.* New York: McKay, 1956.

Dinkheller, A., J. Gaffney, and E. L. Vockell. *The Computer in the Mathematics Curriculum.* Watsonville, Calif.: Mitchell, 1989.

Fuller, K. "Beyond Drill and Practice: Using Computers to Build Hypotheses and Design and Conduct Experiments." *Science and Children* 23 (7) (1986): 9–12.

King, R., and E. L. Vockell. *The Computer in the Language Arts Curriculum.* Watsonville, Calif.: Mitchell, 1991.

McQuade, F. "Interdisciplinary Contours: Art, Earth Science, and Logo." *Science and Children* 24 (1) (1986): 35–37.

Schibeci, R. A. *Computers in Science Teaching: A Select Bibliography, 1983–86.* Project Description, Murdoch University, Australia. ERIC Document ED-287-688, 1987.

Vockell, E. L., and E. Schwartz. *The Computer in the Classroom.* Watsonville, Calif.: Mitchell, 1988.

Vockell, E. L., and R. van Deusen. *The Computer and Higher-Order Thinking Skills.* Watsonville, Calif.: Mitchell, 1989.

Whitaker, B., E. Schwartz, and E. L. Vockell. *The Computer in the Reading Curriculum.* Watsonville, Calif.: Mitchell, 1989.

CHAPTER 5

THE COMPUTER IN THE SECONDARY SCIENCE CLASSROOM

By ENHANCING ACADEMIC LEARNING time (ALT), the computer can help students achieve important objectives in the secondary science classroom. Sometimes computers can help structure the learning experiences more effectively than other media. In other cases, it may be possible to include skills and concepts that could not even be addressed without computers. Using the instructional, emancipatory, revelatory, and conjectural categories employed in preceding chapters, this chapter first discusses the basic ways in which the computer can be used in the secondary science classroom and then examines computer applications to specific areas of the curriculum, including biology, Earth science, chemistry, and physics.

Computer applications in secondary science have the potential to help teachers focus instruction on two essential goals: (1) to help students acquire and practice applying key concepts important to understanding natural phenomena, and (2) to help students develop higher-order thinking skills in science, such as formulating and testing hypotheses, analyzing data, making conclusions, and constructing scientific models.

In the first case, carefully designed tutorial and drill-and-practice software can increase the amount of ALT students spend learning and applying important concepts in science. In the second case, well-designed simulations, microcomputer-based laboratories (MBLs), database/ spreadsheet activities, and telecommunications/networking projects can provide quality ALT on higher-order thinking and problem-solving skills. Positive attitudes toward and about science can also be developed. For example, students can learn to modify concepts in the face of new evidence, to seek data to validate observations or explanations, to view hypotheses as ideas to be tested, and to show respect for the ideas of others.

In addition, tool applications such as word processing, computer-managed instruction programs, test-generating programs, computerized testing programs, and equipment-inventory and supply-tracking programs help make the science teacher more efficient in developing and delivering instruction.

INSTRUCTIONAL USES

Drill-and-practice programs give students feedback for responses as they practice skills or concepts previously taught. Chapter 2 presented five guidelines to help avoid pitfalls in the misuse of computerized drills in science education. As long as these pitfalls are avoided, good drills can provide excellent opportunities for practice and feedback with regard to important scientific concepts, principles, and skills at the secondary level. Some examples of traditional, secondary-level science objectives that often require some drill, practice, or tutorial work include the following:

- Labeling the parts of the cell and explaining the function of each part
- Naming the three types of rocks
- Defining the word *conductor*
- Identifying the constellation Orion
- Listing the parts of an atom
- Identifying an organism as a plant or an animal
- Explaining the relationship between earthquakes and volcanoes
- Showing how a group of eight organisms are related in a food chain/ web

In short, good science drills enable students to apply scientific skills and concepts in a wide variety of settings. Examples are found throughout this chapter.

EMANCIPATORY USES

In secondary-level science education, the role of the computer as a tool can be very important. Science teachers use the power of the microcomputer for the daily procedural management of the classroom—attendance, grades, lesson preparation, and test preparation. Both students and teachers use the word processing capabilities of the computer to ease the drudgery of writing lab reports, and spreadsheets facilitate data transformation and various repetitive mathematical calculations. The computer can draw and redraw graphs in various forms, another important use in the science classroom.

One of the most important objectives in science education, however, should be learning to use the scientific method—that is, problem solving and hypothesis testing, which is the topic of Chapter 6. One critical skill in hypothesis testing is the ability to compile, analyze, and draw conclusions from a set of data. All too often, this data set is very small because of time and budget constraints or because of the inability of the human to make sense from huge quantities of data. The computer's ability to analyze large amounts of data in a short time makes it an ideal tool in many science laboratory experiments. For example, an MBL gives science students the opportunity to see the results of their experiments while they are conducting them; thus, they can analyze data and redesign experiments quickly, many times making adjustments "on the fly."

The microcomputer database (Figure 5.1) offers a useful and instructive tool for the science classroom, allowing students to enter, sort, arrange, select, and organize data based on their own hypotheses. Students can test "what if" questions or employ the database to quickly eliminate many unwanted possibilities. It is important that, along with the database manager and the sample data in the files that accompany it, solid teaching strategies and student materials are provided in a teacher's manual or guide. The well-designed database eliminates the need for hand-written charts and keys, which often waste considerable time by requiring students to flip pages, find the appropriate section on a page, identify choice statements, flip some more pages, and then repeat this process. The computer greatly facilitates this process and enables students to focus on the real task at hand. To further simplify matters, supplementary programs such as PFS: GRAPH (Software Publishing Corporation) and TIMEOUT GRAPH (Beagle Brothers) can easily interface with databases and spreadsheets to provide histograms, bar graphs, pie graphs, and other graphic representations to help interpret data. These and other computerized tools can be effectively integrated with each of the subject areas of science (discussed later in this chapter).

Not only do computerized tools free students from routine and trivial tasks, but they also free teachers to be more creative and productive. For

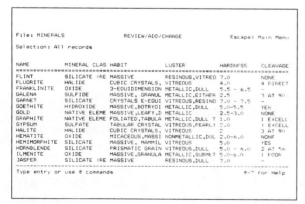

(a)

(b)

Figure 5.1 A mineral database (screen a) that can be sorted and arranged (screen b) to reduce the complexity of the data.

example, the BIOLOGY TEST MAKER (J & S Software) produces tests or quizzes from a bank of 1,500 questions based on up-to-date concepts in high school biology. The teacher can mix questions from different units, randomize question formats, and save tests on disks. Question categories include "Animals and Plants," "Biochemistry," "Diversity," "Ecology," "Genetics," "Living Things," "Physiology," and "Reproduction." Similar TEST MAKERS are available for other areas of science. In addition, many textbook companies have started to produce computerized test-item banks to accompany their textbooks.

Inventory maintenance is the bane of many science teachers. The SCIENCE DEPARTMENT (Condor Computing) maintains an inventory of the high school life or physical science materials in three areas: "Chemicals," "Capital Equipment," and "Equipment and Supplies." Its features include printing of inventory lists, performing searches using multiple search strategies, fast scanning of records, and storing multiple files. Data sorting is quick, printing files in various orders is easy, and adding or deleting entries is simple.

Computerized tools that free science teachers and students for more important tasks are discussed in detail in Chapter 7. MBLs are the topic of Chapter 8.

REVELATORY USES

Simulations at the secondary level should help students learn and apply higher-level science problem-solving skills as well as develop accurate conceptions of life, earth, and physical science phenomena that would be difficult or impossible to study in the high school laboratory. A study by Rivers and Vockell (1987) suggests, in fact, that using computer simulation for problem solving in high school biology has double benefits. It is as good a method as traditional instruction for teaching science facts and concepts and a much better method for teaching problem-solving skills in science. Some evidence was found that the problem-solving strategies were generalized to areas outside science.

Most of the simulations, games, and role-playing situations detailed in this chapter include a teacher's guide with rather detailed background material on the simulation as well as extensive student instructional material. This instructional support is needed if the teacher is to realistically integrate the simulation into the instructional stream. Simulations done only to take up instructional time or as an afterthought do not usually result in enhanced ALT. The teacher must prepare the students to benefit from the simulation, place them in productive learning groups during the simulation, and "debrief" them after the simulation. The simulation should be used in a learning cycle approach, helping students explore the phenomenon under investigation, invent the concept(s) to explain the phenomenon, and apply the concept in a new setting.

CONJECTURAL USES

Conjectural programs provide secondary students with a valuable tool to build robust models of physical, Earth, and life science phenomena. The strength of these programs lies in their ability to supplement and interact with regular science laboratory experiences. Students would probably begin the exploration phase of a learning cycle using regular laboratory materials under teacher direction to develop a preliminary hypothesis. During later stages of the learning cycle (concept introduction and concept application), the conjectural computer program allows students to test their hypotheses quickly, modify them based on the new data, and then go back to the real laboratory to try them out. This cycle, from laboratory to model-building program and back, could occur several times during the practice phase of instruction, until the students' models are complete and generalizable.

The program LAWS OF MOTION (EME), mentioned in Chapter 4, is a good example of a conjectural program that could be used in a secondary physical science class. Students begin testing the movement of a block down an inclined plane in the laboratory, develop a tentative model or "law" for the motion observed on the plane, try out the "law" by manipulating the computer program, modify the "law," and try to confirm their model or "law" in the laboratory.

In this mode, the computer program aids students in several ways: (1) by providing the environment for quickly designing and carrying out simulated experiments without major sources of error, (2) by providing visual support for the development of the scientific model, and (3) by allowing students to "play with" a variety of perspectives and situations that would be difficult to duplicate in the laboratory. For example, LAWS OF MOTION allows the movement of objects down an inclined plane in a frictionless environment.

SECONDARY SCIENCE COMPUTER APPLICATIONS

In this chapter the application of the computer to each of the specific subject areas will be discussed separately. In addition we shall continue to use the instructional, revelatory, conjectural, and emancipatory categories previously described in Chapter 3.

Biology

Biology, or life science, is typically taught at the freshman or sophomore level of high school. It is usually a student's first formal introduction to many of the concepts, terminology, and laboratory skills associated with life science. The texts tend to be encyclopedic, because they need to satisfy so many state adoption guidelines. In fact, there are probably more new vocabulary words in a beginning biology text than there are in an introductory foreign language text.

The tendency in most schools is to try to "cover" this encyclopedia of knowledge without regard to any depth of understanding of key concepts and principles. Emphasis on higher-order thinking skills in biology instruction is lacking because "you don't have time to develop them and most students don't need them" (a quote directly from a practicing biology teacher). As a result, biology laboratories selected by teachers tend to be those that take a minimum of time and that confirm concepts introduced in the classroom discussion or lecture. Very few laboratory activities require students to use higher-order thinking skills to develop concepts or scientific models of the phenomena they are studying.

Finally, much of modern biology that depends on biochemical and biophysical understandings is de-emphasized in introductory courses. Students' background in both chemistry and physics is not sufficiently deep at the freshman level to support such an emphasis. A reorganization of the entire science curriculum at the secondary level must be accomplished to address this issue. A National Science Teachers Association (NSTA) task force, formed in the fall of 1989, is working to provide a framework for such a reorganization. A tentative proposed framework calls for important concepts of biology, chemistry, physics, and Earth science to be taught in appropriate sequences beginning in grade 7 and ending in grade 12 (Table 5.1). Quality time each year would be devoted to teaching core concepts and skills related to each field, with an emphasis on interdisciplinary skills. Study in early grades (7 and 8) would be based on

Table 5.1. The proposed scope and sequence coordination model developed by the NSTA.*

Grade Level	Seven	Eight	Nine	Ten	Eleven	Twelve	
Subject	Hours per Week						Yearly Hours
Biology	1	2	2	3	1	1	360
Chemistry	1	1	2	2	3	2	396
Physics	2	2	1	1	2	3	396
Earth/space science	3	2	2	1	1	1	360
Total hours per week	7	7	7	7	7	7	
Emphasis	Descriptive, phenomeno- logical		Empirical, semiquan- titative		Theoretical, abstract		

* From Aldridge.

observation of concrete phenomena, in the middle grades (9 and 10) based on empirical investigation, and in the upper grades (11 and 12) based on scientific theory.

The Instructional Mode

Good science tutorials have extensive branching and require frequent student interaction. The goal of most of these programs is to develop understanding of complex concepts in physiology, genetics, respiration, photosynthesis, cell structure and function, growth and development, and ecological relationships. Such concepts require that tutorial programs follow the learning cycle, by (1) introducing clear examples and nonexamples of the concept, (2) requiring students to verbally state or recognize the concept, and (3) asking students to apply the concept in a different setting. Diagnostic feedback during Steps 2 and 3 is especially important for accurate concept recognition and concept generalizability.

Tutorial programs that use the learning cycle approach usually have large written and visual databases for Step 1, providing many examples and patterns of attributes. They require a great deal of flexibility and interactivity, including branching, in Steps 2 and 3, so that students receive feedback at crucial points when they are making discriminations and applying the concept. Many tutorial programs have some elements of this approach, but few have been designed with all three steps in mind.

COURSEWARE SERIES IN BIOLOGY (Prentice-Hall) is a good example of a tutorial that uses text with graphics and animation to move students through all three steps of the learning cycle. Branching gives students feedback when the answer they choose is not correct. The program then returns students back to the question. Each tutorial is a menu-driven program that allows students to preview the sequence of objectives for the program, choose instruction related to one or a series of objectives, and then take a test over the same objectives. After each test, the program records responses and reteaches the concept related to each item that was missed. While in the reteach mode, students can access a help menu or a glossary. The record-keeping system holds up to 100 student scores and has print capability. The system enables the teacher to access student scores, either individually or by class. Each test item is correlated to a section in the program, enabling the teacher to see a student's or a class's performance on a specific objective. The tutorial package also includes discussion questions, library investigations, and reproducible practice sheets. The series includes a variety of topics in biology, including reproduction, respiration, and the circulatory system.

Most drill-and-practice programs in biology are designed to help students review the vast array of terminology used in a study of the field (Figure 5.2). These assume that if students can match verbal definitions or labels with the appropriate term or structure, they have accurate knowledge of the concept as well. Focus Media's BIOLOGY KEYWORD SERIES

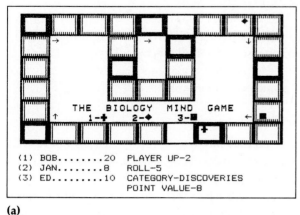

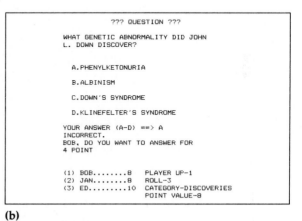

(a)   **(b)**

Figure 5.2 BIOLOGY MIND GAMES helps students review the terminology used in biology classes. Screen (a) shows the game board, and screen (b) displays a question screen from the category "Discoveries."

and GREAT BIOLOGY KNOWLEDGE RACE drill students on knowledge of terms and comprehension of concepts in a game setting. In the BIOLOGY KEYWORD SERIES, secret words are discovered by uncovering clues that include synonyms, examples, definitions, and applications of an extensive library of terms. In the GREAT BIOLOGY KNOWLEDGE RACE, students challenge each other to answer questions drawn from traditional topics in biology courses, such as biochemistry, animal and plant maintenance, life and the environment, and reproduction and development. BIOLOGY MIND GAMES (Diversified Educational Enterprises), designed for content review of general biology, contains 140 questions in four categories: "Biologists," "Discoveries," "Terms," and "Organisms." Students compete with each other to move a token around a game board on the screen. The teacher can edit questions, add questions, and delete and add categories—a useful feature. Besides supporting classroom learning, these programs could be used to prepare students for "Science Bowl" or "Science Olympiad" competitions.

VISIFROG and ANATOMY OF A FISH, two programs from Ventura, use both visual and text databases to help students practice identifying the names and functions of animal structures. Various games help students identify structures, spell terms, and match structures with function and with terminology.

Several recent direct instructional programs reflect a concern for the application of biological research to social and personal decision making (bioethics). These include GENETIC ENGINEERING (Helix), AIDS: THE NEW EPIDEMIC and AIDS: THE INVESTIGATION (Marshware), and SEXUALLY TRANSMITTED DISEASES (HRM). These programs can help the teacher provide an opportunity for students to make relevant application of biological principles in personal situations that they currently face or that they may face at some point in their lives.

The Emancipatory Mode

The computer can serve as a tool in biology classes to release students and teachers from routine, nonproductive work in nearly all the ways described earlier. MBLs (see Chapter 8) can be especially effective tools in studying human physiology and behavior. Such probeware gives students immediate feedback on the progress of their experimentation, which is crucial to the development of accurate concepts and skills in designing and carrying out experiments.

For example, HRM has applied MBL strategies to biology by designing a series of probeware programs in human physiology for high school biology. With BODY ELECTRIC, for example, students can accurately monitor and measure brain waves, produce electrocardiograms, and record the electrical activity of muscles. Using the computer as a tool, students record and analyze data from their own experiments in real time.

EXPERIMENTS IN HUMAN PHYSIOLOGY (HRM) provides accurate detectors of heart rate, respiration rate, skin temperature, and response time, which allow students to record their own and others' physical reactions in real time directly on the computer display. In recording the same phenomena over a period of time, students can use the computer to create graphs of physical behavior (Figure 5.3). Students can investigate topics like homeostasis, biofeedback, the physiology of exercise, stress, and sleep.

Database management programs can also play a useful role in secondary biology classes. DATAQUEST: NORTH AMERICAN MAMMALS (MECC) is a database bank filled with physical, geographical, and behavioral information about ninety-eight mammals native to North America. Secondary students can effectively use ANIMAL LIFE DATABASES and the ENDANGERED SPECIES DATABASE, two Sunburst life science databases used with the BANK STREET SCHOOL FILER. The former contains information on the physical characteristics, classifications, behavior, food,

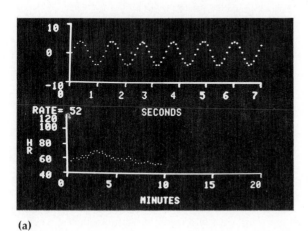

(a)

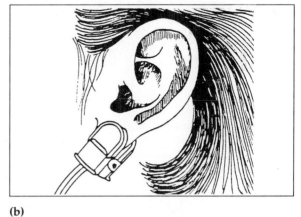

(b)

Figure 5.3 Screen (a) shows a graph displaying the results of a heart-rate experiment from EXPERIMENTS IN HUMAN PHYSIOLOGY. Figure (b) shows the attachment of the clip-on sensor used to measure heart rate.

habitat, reproduction, and adaptations of warm- and cold-blooded vertebrates. The latter contains files on all U.S. endangered mammals, extinct
animals from 1600 to the present, and some of the most critically endangered species worldwide.

Computerized dichotomous keys are specialized databases that allow
students to identify organisms as members of specific taxonomic groups.
As students move through the program, they make choices based on the

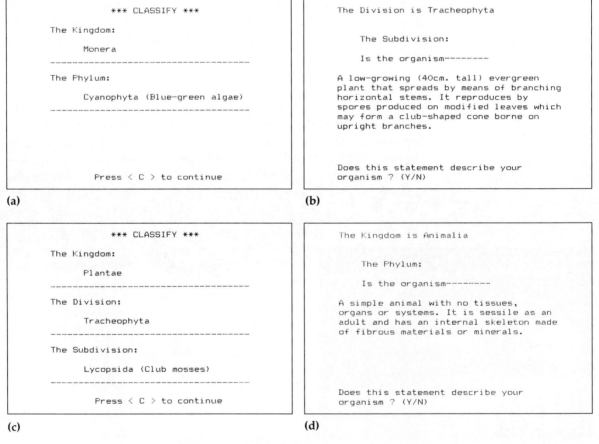

(a)

(b)

(c)

(d)

Figure 5.4 CLASSIFY presents a set of computerized dichotomous keys. Screen
(a) shows the final identification of blue-green algae; screen (b) shows one of the
questions in the dichotomous key and the result; screen (c) shows the final
identification of the club moss; and screen (d) shows one of the questions in the
Animalia branch.

characteristics of the organism they wish to identify. Students make one out of two choices at each point in the program, finally narrowing their choice down to the correct category for the organism. CLASSIFY (Diversified Educational Enterprises) uses a five-kingdom classification system, which allows students to identify the phylum or class to which an organism belongs (Figure 5.4). DICHOTOMOUS KEY TO POND MICROLIFE (EduTech) allows users to identify seventy-nine of the most common protist and cyanobacteria in ponds, puddles, or similar locales; in hay infusions; or in commercially cultured material from biological supply houses. Students can also query the database directly to learn which microlife forms can be expected in a given microhabitat or combination of habitats and which exhibit a particular behavior or condition. With the glossary, users review definitions of terms that appear in the key either independently or when using the key.

Biology students can also use the telecommunications capability of the computer to be a part of the TERC Star Schools Project described in Chapter 4. Any senior high school committed to improving its math and science programs is eligible to apply for participation in the project. Schools from districts eligible for Chapter 1 funding for disadvantaged students are especially encouraged to apply.

Besides tools for biology students, there are also useful tools for biology teachers. For example, along with the more general tools described earlier, the teacher can use GENETICS PROBLEM SHOP (EME) to generate and print an almost unlimited variety of problems, from a choice of thirty different types. After setting the difficulty level, the teacher selects problem values or has the computer select them. It features monohybrid, dihybrid, and trihybrid crosses; human or plant traits; dominant and recessive alleles; word or symbolic form problems; Punnett square problems; and genotypic and phenotypic ratios.

The Revelatory Mode

Simulations in biology have been developed in areas where the lack of technical skill, supplies, or time make it almost impossible to provide live laboratory experiences. Many of these simulations are designed to be substitutes for the live laboratory. Some are designed to introduce or follow up biology laboratory experiences. One popular topic of simulation programs is genetics. These programs compress time, so that many generations of organisms can be "raised" within a short period of time, helping students to design and carry out genetics experiments involving many generations within one or two class periods. One of the most useful of these is DESIGNER GENES (QED) (Figure 5.5). This program combines a tutorial on basic Mendelian genetics with computer-generated monohybrid or dihybrid crosses of simple alleles, multiple alleles, and sex-linked alleles.

```
        DESIGNER GENES

MAIN PROGRAM MENU

  1. INTRODUCTION

  2. GENETICS TUTORIAL

  3. PROBABILITY TUTORIAL

  4. SIMPLE ALLELES

  5. MULTIPLE ALLELES

  6. SEX-LINKED ALLELES

  7. END THIS SESSION

Press the NUMBER of the section
that you would like to study.
```

(a)

```
     A victim of SICKLE-CELL ANEMIA has red
blood cells that elongate into a thin
'sickle' shape.  This affects the
blood's ability to carry oxygen to the
body.  Symptoms of the disease include
dizziness, shortness of breath, and
joint and muscle aches.  The disease
can be fatal.

     Sickle-cell anemia is a hereditary
disease that results from the presence
of two recessive genes.  A heterozygous
person will not show any symptoms of the
disease.  The symbols used for the
sickle-cell alleles are:

     A for NORMAL, dominant
     a for ANEMIC, recessive

        Press > to continue
```

(b)

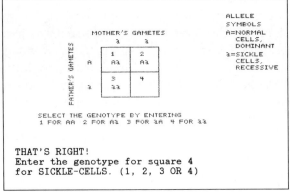

(c)

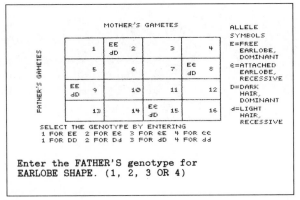

(d)

Figure 5.5 DESIGNER GENES is an introduction to the study of heredity. The program contains a tutorial on basic genetics and a simulation that allows students to select and study a variety of monohybrid and dihybrid crosses. Genotypes of possible offspring are computer-drawn, and a Punnett square is drawn. The student is asked to fill in the missing genotypes. Screen (a) shows the main menu. Screen (b) shows an introduction to a monohybrid cross involving sickle cell anemia. Screen (c) shows the successful filling in of the Punnett square for the sickle cell cross, and screen (d) shows the Punnett square for a dihybrid cross involving earlobe shape and hair color.

After the student chooses from a list of several genes, the computer generates the cross, draws a Punnett square, and asks the student to provide the missing genotypes. A similar program is MENDELBUGS (Focus Media), where students can select parent bugs and trace the traits through monohybrid and dihybrid crosses. They can predict the types of offspring and compare their predictions with the actual outcomes. Similar genetic programs include MONOCROSS and DICROSS (Diversified Educational Enterprises).

Two well-know genetics programs written by Judith Kinnear are BIRDBREED (EduTech) and HEREDITY DOG (HRM) (Figure 5.6). BIRDBREED simulates the inheritance of color in parakeets. Crosses, performed between any two birds within a breeding group, illustrate the phenomena of dominance, sex linkage, independent assortment, multiple alleles, gene interaction, and linkage between genes. Four levels of difficulty enable the program to be used in both beginning and advanced biology. HEREDITY DOG allows students to mate dogs of different coat colors and patterns while the computer generates pups. Experiments involving both single-gene or two-gene inheritance can be done. CATLAB (Conduit), a similar program, allows students to mate domestic cats on the basis of coat color and pattern.

HUMAN GENETIC DISORDERS (HRM) explores Huntington's disease, cystic fibrosis, hemophilia, Tay–Sachs disease, albinism, and Duchenne's muscular dystrophy by having students make inferences about the parents' genotypes and phenotypes or through a computer-generated pedigree and pattern of inheritance.

Another popular topic for simulations in biology is ecology. This area is popular because of the difficulty in actually observing many ecological

Figure 5.6 This screen from HEREDITY DOG shows the mating of two dogs with similar coat patterns. Observing the offspring helps students develop the concepts of allele segregation and gene assortment.

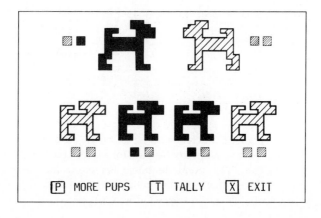

phenomena, most of which occur over long periods of time. POLLUTE (Diversified Educational Enterprises) is a student-interactive simulation that examines the impact of various pollutants on typical bodies of water (Figure 5.7). The laboratory guide directs students to study the impact of water temperature, waste type, waste-treatment type, rate of waste dumping, and type of receiving body of water on the oxygen content of the water and the survival of fish living in the water. The computer generates tabular and graphic data to enable students to test various hypotheses. The teacher's

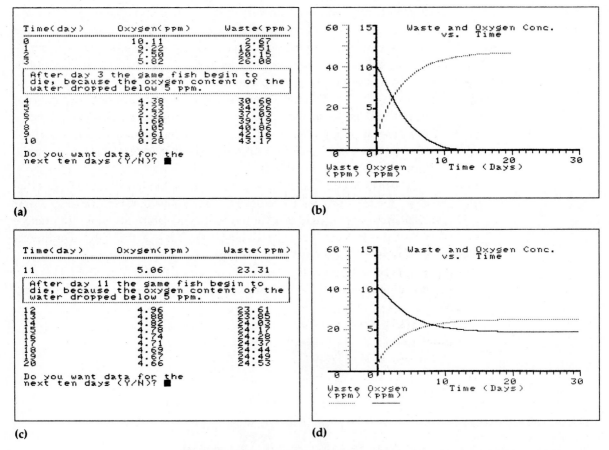

(a)   (b)   (c)   (d)

Figure 5.7 Screens from POLLUTE show the (a) tabular and (b) graphical representations of the data for a large lake, at 45° F, with industrial sewage at 11 ppm/day and no treatment. Screens (c and d) show the same experiment, but the sewage has undergone primary treatment.

guide contains background on the simulation, problem-solving objectives, a test keyed to the objectives, and an answer key to the student laboratory guide and the test.

BALANCE (Diversified Educational Enterprises) explores the variables affecting predator/prey relationships (Figure 5.8). Students manipulate variables: type of environment, number of deer, number of wolves, and deaths by natural and unnatural causes. Tabular and graphical output illustrates the effects of the variables on the chosen population. The teacher's guide contains background on the simulation (see the box on pages 118–119), problem-solving objectives, a test keyed to the objectives, and an answer key to the student laboratory guide and the test.

NICHE (Diversified Educational Enterprises) is an ecological game/simulation that requires students to correctly place one of five organisms in its proper ecological niche by specifying the environment, range, and competitor for the organism. The organism flourishes or fails depending on the selections. The user's guide features background information, suggestions for use, and analyzed sample runs of the program.

Another important area for biological simulations is population genetics and evolution. Included here is MOTHS (Diversified Educational Enterprises), in which students observe the peppered moth case in industrial England. In this simulation, students manipulate the numbers of light- and dark-colored moths in different environments to determine the change in frequency of the genes for light and dark wing color over a number of generations. Included are a teacher's guide and student laboratory manual. SURVIVAL OF THE FITTEST (EME) uses a game format. Students must

Figure 5.8 This screen shows the pattern of deer (thick line) and wolf (thin line) that would emerge from student-selected variables in BALANCE.

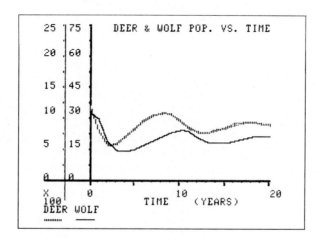

Balance: Lesson Based on a Problem-Centered Learning Cycle

This activity is based on BALANCE, a high school life science simulation described elsewhere in this chapter. The objective of the lesson is, given a simulated problem and data pertinent to that problem, students will be able to identify inferences that are consistent with the data provided and that describe the patterns of change in the predator and prey populations, which emerge over time as changes in the environment, numbers of predator, and numbers of prey occur.

Teacher's Guide

1. Introduce Problem A (from the student laboratory manual) to students. The problem asks students to determine the effect of different environments on portions of a predator/prey population and to suggest possible reasons for any population size differences due to the environment.
2. Using the computer simulation in a whole-class demonstration mode, read through the guidelines for running the simulation. Ask students to suggest specific experiments that may provide data to help solve the problem. Using students' suggestions, make several runs through the simulation, collecting data and discussing the implications. If students do not suggest a crucial experiment, suggest that additional data are needed and ask a student to suggest an experiment to collect the missing data.
3. Have students describe the patterns of change they see in the predator/prey populations. Ask students for hypotheses that could explain the population changes. Divide the class into small groups; ask each group to develop an experiment to test a hypothesis of their own choosing. Instruct groups to collect data related to their hypotheses and analyze the data to determine if their hypotheses are either confirmed or rejected.
4. Have each group report on their experiment. Discuss the patterns of population change observed. Ask students to report whether their hypotheses are confirmed or rejected. Help the class compare the results of the various experiments, so that a solution or solutions to the problem can be agreed on. Use the terms *carrying capacity*, *growth rate*, *maintenance level*, and *predation efficiency* to describe relevant concepts as you focus in on the solution.
5. Introduce Problem C to your students. This problem deals with the opening of a new camp site in a national park. Divide the class into small, heterogeneous groups of three to five students. Give each group a worksheet that describes the problem and lists the questions that must be answered to solve the problem. Ask groups to design a series of experiments to solve the problem. Tell them that you want to go over

(Continued)

their plans before they go to the computer simulation to collect data. After making suggestions, allow them to carry out their proposed experiments.

6. Ask each group to turn in the worksheet describing their solution to the problem, with accompanying rationale. Discuss several of them with the whole class, identifying strengths and weaknesses in the designs and strategies used.

7. Make sure that the limitations and assumptions on which the computer simulation is built are made clear to students by this point. If they have not been discussed in Steps 1 through 6, end the lesson with a discussion of the simulation model's limitations and the assumptions on which it is based. These are clearly outlined in the teacher's guide.

Evaluation

Student performance on Problem C could be used to evaluate attainment of the lesson objective. Several test items that measure the attainment of the lesson objective are provided in the teacher's guide.

capture as many prey as possible by directing the evolution of the predator while the computer guides the prey's evolution. It includes a record-keeping feature and lab exercises.

PLANT (Diversified Educational Enterprises) simulates the growth of a bean plant. It combines a computer simulation with biology laboratory experiences to give students accurate concepts about plant growth, the ability to construct and interpret graphs, and experience in the application of scientific experimental design.

OSMO (Diversified Educational Enterprises) introduces the student to the process of osmosis in red blood cells, combining a biology laboratory with a computer simulation, where students place a blood cell in solutions of differing salt concentrations. The package includes a teacher's guide and student laboratory section.

MICROBE (Synergistic Software) is a unique adventure game that requires students to work together in differing roles to be successful. Students must plan and execute a complex mission in an environment that is both familiar and alien: the human body. This game is roughly based on the movie "The Fantastic Voyage." The students are in charge of a medical research submarine that carries a crew of four: a captain, a navigator, a technician, and a physician. Their job is to determine the cause of illness and effect a cure of one of a number of patients, who are cryogenically frozen. Besides learning and applying problem-solving strategies, students learn about human anatomy, immunology, the effects of brain damage, drugs and their uses, hemorrhages, and clots, as well as infectious bacteria, viruses, parasites, and fungi that may invade the body.

The Conjectural Mode

In this category, computers are used to build interactive models of various science phenomena, so that hypotheses or configurations can be formulated and tested. Computer applications in this category are *model-building programs* and *microworlds*.

Model-Building Programs. OSMOSIS AND DIFFUSION (EME) helps students build and manipulate a scientific model for the process of osmosis. In the "what if" mode, students identify and control several factors that influence movement of substances into and out of the cell, including temperature, concentration, solubility, molecule size and charge, and membrane pore size. This program can be used to good advantage as students move back and forth from the "wet lab" where they observe the phenomena to the computer model for the process on a molecular scale.

A program that assists students in visualizing the complex biological structure of genetic material is DNA—THE MASTER MOLECULE: LEVEL I: THE BASICS (EME). This program simulates the building of a DNA molecule, the transcription of mRNA, protein synthesis, and the effects of mutation on nuclear material. Students can use this program in conjunction with laboratories using living materials so that molecular models are tied to the process of development and reproduction of organisms.

Microworlds. PLANT GROWTH SIMULATOR (Focus Media) is a microworld of plants, where students can design their own plant-growth experiments. They choose the number of plants to grow, the length of the growing season, and eight growth variables, which include wavelength of light, moisture level, percentage of carbon dioxide in the atmosphere, and the amount of nitrates. This program can be used effectively in conjunction with real plant-growth experiments in the high school science laboratory.

Earth Science

Many states have changed their graduation requirements to require two years of laboratory science, and Earth science is one of those courses added to the curriculum as an acceptable laboratory science course. Several states require Earth science at a specific grade level; many other states have added

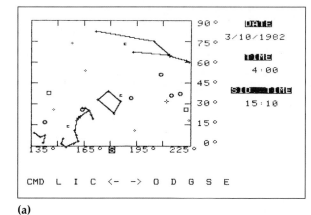

(a)

Figure 5.9 Screen (a) displays the view from TELLSTAR program of the "Grand Conjunction of 1982." Screen (b) shows the "Star Atlas View" of the occultation of Jupiter on August 18, 1990, from Voyager, version 1.2.

Star Atlas View

Chart Center:	RA: 09h 00.0■	Dec: 20° 57'	Field: 83° x 57°
Universal Time:	14:38	08/18/1990	Julian Day: 2448122
Local Mean Time:	06:38 AM	08/18/1990	Epoch: 2000
Observing Location:	122° 09' W	37° 43' N	

(b)

a rigorous Earth science class or specialized Earth science classes such as astronomy, meteorology, oceanography, planetary geology, and space science. The National Science Foundation (NSF) funds programs that are currently defining various aspects of Earth science; it is one of the areas covered by the ETS Advanced Placement tests; a national Earth science test has been developed; and Earth science teachers together with scientists and science educators are currently writing curriculum materials for all grade levels. Software developers and Earth science teachers have been quick to recognize the usefulness of the computer in the K–12 Earth science classroom.

The Instructional Mode

Earth science programs in astronomy center around finding and identifying the sun, moon, planets, stars, constellations, comets, and deep-sky objects. Many programs exist to turn the computer into a planetarium, allowing varying degrees of sophistication. TELLSTAR: COMET HALLEY EDITION (Spectrum HoloByte) puts such a planetarium on the computer screen (Figure 5.9). Students can program their own date, time, and location and can save the last view to disk and print it. All types of objects can be identified, and patterns between constellations or the rising and setting of sky objects at different times in the year are obvious. VOYAGER (version 1.2) (Carina Software) provides an interactive desktop planetarium on the

Macintosh computer. Besides viewing the sky and all its objects in the simulated sky on any particular evening, users can track a planet across the sky or over several years, change viewpoints (e.g., move the point of observation to Mars), and zoom toward and away from objects to investigate relationships and apparent movement. ASTRONOMY: STARS FOR ALL SEASONS (Educational Activities) (Figure 5.10) includes only the major objects that can be seen from the backyard and reduces the complexity for users. A problem with this program, however, is that it introduces and uses the terms *right ascension* and *declination* without sufficient explanation. Using the terms *altitude* and *azimuth* would be more appropriate for general users not particularly interested in the complexity of astronomy. (This is not an insurmountable problem, provided the teacher is willing to teach these terms in a context outside the computer program.) SOLARISM (Interstel), a program for IBM and compatibles, has a simulation of Comet Halley's orbit around the sun and has sky scenes similar to those in TELLSTAR.

Many teachers have found model rocketry to be a motivating way to introduce learners to the Earth sciences, and students often enjoy model rocketry as an academic hobby. Several programs produced by Estes Industries introduce students to the building and flight of model rockets. Students can learn about the parts of a rocket, types of rocket engines, and forces acting on rockets; or they can enter specifications for a rocket they would like to build and predict various flight parameters.

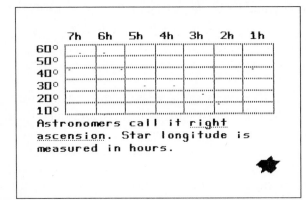

(a)

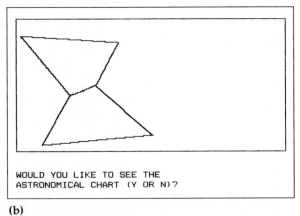

(b)

Figure 5.10 Screen (a) from the introduction of ASTRONOMY: STARS FOR ALL SEASONS helps define the term *right ascension*. Screen (b) shows the constellation Orion with lines drawn in to highlight the constellation.

Other drill and tutorial programs for the Earth sciences include explanations of the rock cycle, rock and mineral identification, and meteorological programs. For example, WEATHER FRONTS (Diversified Educational Enterprises) provides an excellent presentation on weather map symbols, weather maps, and weather prediction. Either the whole class or individual students can use it. The program documentation offers one of the best explanations of this topic that we have seen and should be used during the concept introduction and application phases of the learning cycle. After studying the tutorial explanation, students are given weather information and asked to predict the weather one day ahead. If students miss the prediction, the computer demonstrates a correct way to think through the problem.

The Emancipatory Mode

Every Earth science textbook (and many elementary textbooks containing Earth science concepts) includes at least one chapter on the identification of rocks and minerals. Identifying Earth materials is usually done in a very traditional method, which has been passed down from college to high school to middle school to elementary school. In this method, the properties of each mineral (or rock) are presented to be memorized, and, perhaps, students test these properties in the laboratory to verify what has been learned from the book.

For example, limestone or calcite reacts with hydrochloric acid (HCl) by bubbling and releasing carbon dioxide (CO_2). Students are given a piece of limestone and some weak HCl and observe the bubbling. Students may be allowed to test a few other rocks or minerals to make sure that they don't also react to the acid. During the test over this material, students may not see any rocks or acid but simply be asked a written question: "Name the rock that reacts with HCl."

In contrast, there is a great body of literature collected over many years regarding the use of "discovery" or "inquiry" learning in the science classroom as opposed to "expository" learning (Wise and Okey, 1983). This approach encourages teaching science as a process and not simply the set of conclusions formed by previous generations of scientists. Few topics in science call for the discovery method as obviously as does mineral and rock classification. Most teachers, however, continue with the rote approach because it is very time-consuming and inconvenient to conduct the series of experiments needed to discover an effective classification system.

In learning to observe, classify, and identify Earth materials, students must actively employ concepts that they themselves develop. Students must generate their own concept label, define that concept and identify the criterion attribute(s), relevant attributes, irrelevant attributes, examples, and nonexamples. Further, students must identify relationships between two concepts, such as metallic luster and specific gravity. Once that has been done, students attach the name by which that mineral is known to others.

The microcomputer provides the possibility of great advances in our ability to use the inquiry approach. For instance, by employing the Sort Records, Find Records, and Arrange features of a database, students can quickly eliminate many of the characteristics that do not apply and discover one of several possibilities for the sample (Figure 5.11). In addition, students learn which characteristics or attributes are the most relevant or critical for identifying materials and establishing a classification system.

The teacher or students can construct a mineral database file from illustrations or keys in their textbook, from a layperson's guide to identification of rocks and minerals, or from a college or university textbook. It will be necessary to construct separate files for rocks, minerals, and fossils because the key criteria for identification are different for each.

The minerals database shown in Figure 5.11 uses APPLEWORKS and contains 127 records. Each record contains sixteen categories useful in identifying minerals. The accompanying word processing file explains the

```
File:    MINERALS                                              Page  1
Report: METALS                                                 8/1/90
Selection: LUSTER contains METALLIC
   and      HARDNESS is less than 3.0
   and      SPECIFIC GRAVITY is greater than 7.5
NAME          LUSTER            HARDNESS    SPECIFIC GRAVITY  STREAK
------------  ----------------  ----------  ----------------  ----------
CINNABAR      ADAMANTINE,SUBME  2.0-2.5     8.1               SCARLET
COPPER        METALLIC          2.5-3.0     8.9               ORANGE-RED
GALENA        METALLIC,EITHER   2.5         7.6               GRAY-BLACK
GOLD          METALLIC          2.5-3.0     19.3              YELLOW
SILVER        METALLIC          2.5-3.0     10-11             GRAY-WHITE
```

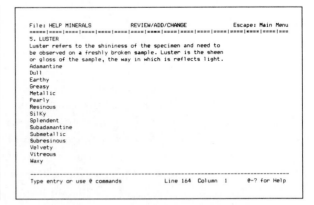

Figure 5.11 Students can use this mineral database to eliminate all minerals except a few, so that other characteristic properties can be viewed and then tested.

```
File: HELP MINERALS           REVIEW/ADD/CHANGE           Escape: Main Menu
=====|====|====|====|====|====|====|====|====|====|====|====|====|====|====|===
5. LUSTER
Luster refers to the shininess of the specimen and need to
be observed on a freshly broken sample. Luster is the sheen
or gloss of the sample, the way in which is reflects light.
Adamantine
Dull
Earthy
Greasy
Metallic
Pearly
Resinous
SilKy
Splendent
Subadamantine
Submetallic
Subresinous
Velvety
Vitreous
Waxy

-------------------------------------------------------------------------------
Type entry or use @ commands              Line 164  Column  1      @-? for Help
```

Figure 5.12 The word processing file accompanying the database explains how to use the categories, showing the descriptions used in the "Luster" category.

database file, defines each category, and explains the mineral property. The word processing file lists choices for properties such as habit, luster, and mineral class, which can be entered into the database (Figure 5.12).

This unit provides students with practice on the properties of minerals as well as lessons in using the computer and accessing the database and word processing files. After they have practiced using the database, students are given a set of actual mineral samples to identify by name. Students perform whatever tests they wish and use the database to find the name of the mineral. Students can use additional properties in the record of each mineral to find out more about properties that were not tested.

In this manner, students take responsibility for setting up their own investigations, collecting and analyzing their own data, and learning to use a tool that is applicable in other classes and in the world outside the classroom.

ROCKY: THE MINERAL IDENTIFICATION PROGRAM (Michigan State University) is an interactive, inquiry-oriented laboratory experience. This program can easily be used either for exploration or for concept application. It allows users to identify one of fifty-eight minerals by examining the physical properties of color, luster, shape, specific gravity, and hardness. Students select one of the minerals, decide which property should be tested first, and enter the results of the test. The computer provides a list of minerals that have *not* been eliminated as a result of this first test. Students continue this process, eventually eliminating all but one mineral. If students cannot eliminate all but one, the program displays a list of the remaining minerals with their physical properties to help identify the mineral. The purpose of the program is to help students discover which attributes are the most relevant for categorizing each mineral. Many printed and computerized dichotomous keys force a predetermined—and sometimes inappropriate—choice on students.

A similar commercial program is MINERAL IDENTIFICATION COMPUTER PROGRAM (Scott Resources). It comes with four sets of rocks and minerals and is menu-driven. It asks specific questions about each sample, and students must test each item to know its properties. The program tells students if they do not get the properties matched to the right sample. This is a predetermined key and is representative of the textbooks from which it is drawn.

GEOWORLD (Tom Snyder Productions) is a large database that is used to discover the worldwide distribution of fifteen different mineral resources (Figure 5.13). While using this database, students explore for

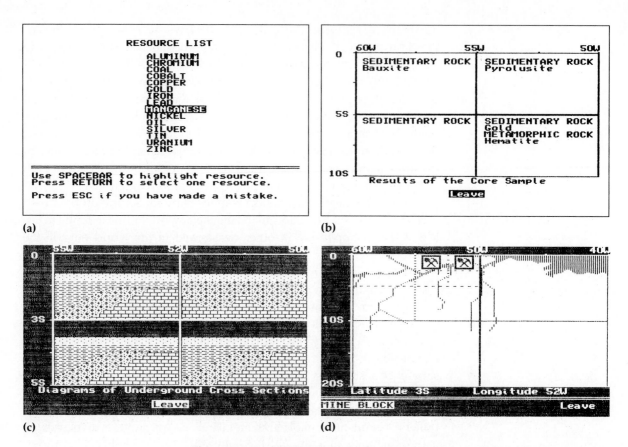

(a)

(b)

(c)

(d)

Figure 5.13 In GEOWORLD, the distribution of fifteen different minerals can be explored (screen a). Screen (b) shows the result of a core sample taken in South America; screen (c) shows cross-sectional diagrams of the upper right quadrant of screen (b); screen (d) shows the mining of the blocks. Manganese was found in the upper right quadrant of screen (c).

minerals or fossil fuels and learn about sedimentary, igneous, and metamorphic rocks. They must use processes of geological exploration and geochemical testing in their search. This database can be used in many ways. For example, students can analyze the worldwide distribution of manganese, determine which state has the most coal, or find out which country has the most gold or aluminum. The database does not contain answers to these questions but permits students to find the answers.

Students can copy the information obtained in GEOWORLD into APPLEWORKS and can create new databases, spreadsheets, or word processing files. In this way, students create reports and keep track of their explorations. They also create new databases that organize the information in a way that is useful. The manual not only explains how to do all of this but also contains lesson plans and other teaching ideas.

The Revelatory Mode

GEOLOGICAL HISTORY (Sunburst) and GEOLOGY IN ACTION (HRM) allow students to construct or study geological cross sections to determine Earth history (Figure 5.14). Students can work backward from the cross section to discover the history. In running the program, students must develop and use a model. Like geologists, they must analyze data and make predictions. GEOLOGISTS AT WORK (Sunburst) allows the creation of cross sections, three-dimensional models, and geological maps from data gathered through the simulated drilling of core samples. For more advanced students of geological cross sections, GEOSTRUCTURES (Intellination) for the Macintosh shows the cross sections in three-dimensional block diagrams, instead of the flat two-dimensional drawings.

Computer-assisted instruction should be built around the ability of the computer to do things that cannot be done in other ways. PLATE TECTONICS (Educational Images), a program for the Macintosh, does just that. Users can move the crustal plates, of which the earth is constructed, both forward and backward in time to generate hypotheses and to test assumptions. This program allows continents to be moved together or apart along different axes, to try to analyze earth movements in the future or to match up areas of continents that were connected millions of years ago.

The Conjectural Mode

Several simulations are designed to be used in a conjectural mode, with students investigating a problem, forming hypotheses, gathering data, and drawing conclusions. For example, in GEOLOGY SEARCH (McGraw-Hill), student teams form oil exploration companies and explore the island of Newlandia for oil. The teams begin by performing a trial-and-error exploration to see how the program works. To use the program effectively, the teams must use the "Searchbook" and library material to learn about oil formation, rock formations, geological exploration, and the interpretation of data. During the exploration and concept introduction phases, the

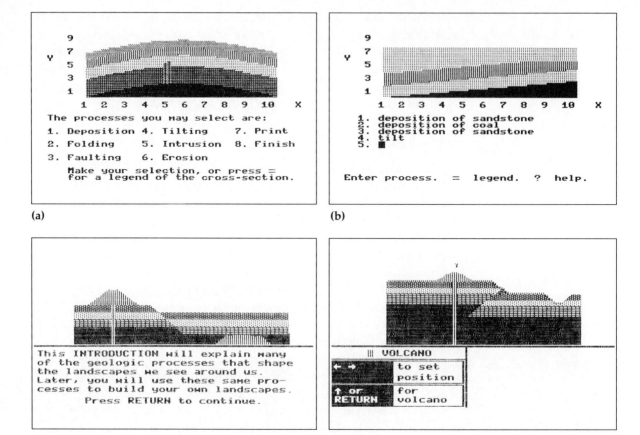

(a)

(b)

(c)

(d)

Figure 5.14 Screen (a) shows a cross section of four layers of sedimentary rock, which have been intruded and folded using GEOLOGICAL HISTORY. Screen (b) interprets a cross section; screen (c) is an example from the tutorial GEOLOGY IN ACTION; screen (d) shows the building of a cross section using GEOLOGY IN ACTION.

teacher can lead the class in discussions to share the information that individuals or the teams have developed. During the concept application phase, teams run various geological tests—density scanning with a gravimeter, core sampling, examining rocks and fossils, seismic surveying, and drilling for oil—to determine the chance of finding oil in a selected geological region. Teams make money by striking oil or natural gas and selling the minerals. The program keeps track of which sections have been drilled and whether minerals were found. The teams must make additional decisions such as whether to sell the oil immediately or to wait and hope for higher prices. The program works well as part of cooperative learning projects (described in Chapter 9), but it can also be used by individual students.

Chemistry

Chemistry is offered in secondary schools as an advanced laboratory science taken primarily by college-bound students. Although textbooks written in the 1960s were inquiry-oriented, demands of content coverage and state assessments have often caused the high school chemistry course to evolve into a shorter version of general college chemistry. High school chemistry texts typically require that students operate at the level of intellectual sophistication described by Jean Piaget as "formal operational." This means that they must deal heavily with abstractions, focusing on things that "might be" rather than on their own concrete experiences. However, investigators have estimated that 50 percent or more of college nonscience majors cannot think consistently in a formal operational mode (Herron, 1975). Instead, a great number of college students (and even more high school students) operate at a concrete level, employing sensory observations and objects in their immediate presence or personal experience in order to think about science. In addition, even when students are *capable* of formal operational thought, they often prefer to have initial concrete experiences on which to build their abstractions. Beginning lessons with concrete activities or concepts and then gradually progressing to more abstract concepts not only makes chemistry accessible to more students but also helps lead them into formal operational thinking patterns. The computer programs described in this section can help supply these concrete experiences.

The Instructional Mode

Since microcomputers first appeared in the science classroom, chemistry teachers have been using them to provide drill and practice and tutorial work in general chemistry. Beginning with the elements and their symbols and ending with organic chemistry and nuclear chemistry, commercial or public domain programs exist for all topics and for all budgets. A serious problem, however, is that some of these programs do not employ good pedagogy. For example, many give the impression that chemistry is difficult and abstract and that the best way to learn it is through memorization without understanding the concepts. This is, of course, a reflection of the textbooks around which the programs are based. There are programs, however, that present this material in a practice mode, asking questions that cause students to think about why the answers are the way they are.

CHEMISTRY: THE PERIODIC TABLE (MECC) and CHEMAID (Ventura) are two databases that help students gain insight into the regular patterns that the elements exhibit. Both programs have a database at their core, which includes basic facts about various elements, including their family and their position on the periodic table. The MECC program (Figure

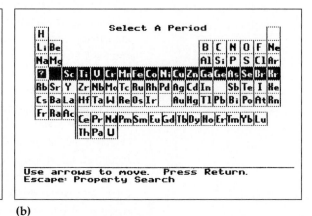

(a)

(b)

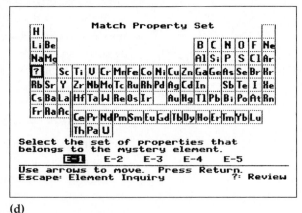

(c)

(d)

Figure 5.15 With CHEMISTRY: THE PERIODIC TABLE, students try to discover the identity of an unknown element (screen a) by selecting other elements (screen b) and viewing specific properties (screen c). When students have enough information, they try to match the element to the correct answer (screen d).

5.15) also provides an unguided property search of the ninety-two natural elements, allowing students to select and compare any elements. The "Element Inquiry" challenges the students to apply their knowledge of the periodic table. This program also allows the prediction of how the elements will react.

CHEMISTRY ACCORDING TO ROF (ROF) offers seventeen diskettes with seventy-nine programs. These programs aren't as "slick" as many from larger commercial companies, but they have something for nearly every topic and the approach shows an understanding of what learning chemistry is all about. For example, while many programs simply drill on the symbols of elements and formulas, ROF presents "rules" on writing formulas and asks why certain formulas are incorrect (Figure 5.16). ROF presents problems in many different ways and always insists on proper use of significant figures. Students use various keys for uppercase and lowercase letters as well as for subscripts and superscripts. These keys are the same in all the programs and are simple to learn.

Although most ROF programs are written for review, the stoichiometry section and some others provide excellent explanations and can be picked up by students who are not familiar with the procedures. The ROF Millikan Oil Drop Experiment and the pH plotting program on the "Acids, Bases, and Salts 2" disk are among those that are suitable to use with the entire class on a large-screen monitor (Figure 5.17). OIL DROP (EduTech) provides another good simulation on the same topic.

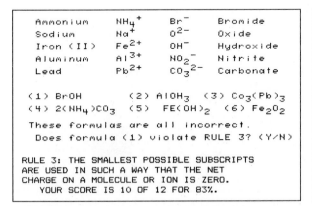

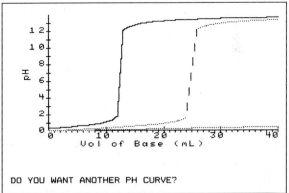

Figure 5.16 In this screen from CHEMISTRY ACCORDING TO ROF, students must decide if the rules for writing formulas have been followed correctly.

Figure 5.17 In CHEMISTRY ACCORDING TO ROF: "Acids, Bases, and Salts 2," users select the concentrations and volumes and observe the resultant graph in an acid/base titration.

Project SERAPHIM, an NSF-sponsored clearinghouse for chemistry software, also has software for nearly every topic covered in high school chemistry. Besides interfacing software (described in Chapter 8), the project maintains and adds to a large library of software, most of which comes with documentation. Project SERAPHIM produces aids that help the teacher select and use SERAPHIM software appropriate for the subjects included in introductory courses. Teacher's guides are keyed to popular introductory chemistry textbooks so that appropriate software can be selected for each chapter. In addition, teaching tips provide detailed information about using various individual programs.

COMPress is a third source of many different programs covering all chemistry topics. COMPress disks offer user-friendly help by indicating typing errors or word omissions and by allowing users to back up one screen or ask for the answer. The format of these disks is mostly drill with some instruction, followed by some practice of the concepts involved. The disks about the gas laws also contain some simulations that allow students to change the pressure or volume to see what effects this would have on the system.

The COMPress ORGANIC QUALITATIVE ANALYSIS is a demonstration of an experiment on organic reagents without the odiferous side effects. Students run one of several standard tests (Figure 5.18) and observe a graphic on the screen showing how it is done. These tests can be run on

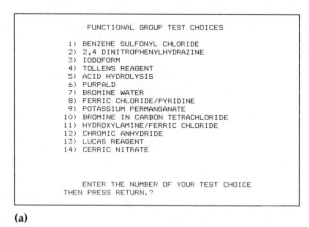

(a)

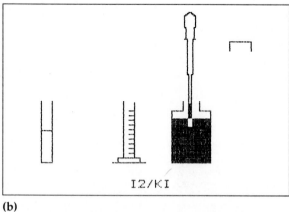

(b)

Figure 5.18 In a simulated experiment in ORGANIC QUALITATIVE ANALYSIS, students first select one of several standard tests (screen a) and then add 2 mL of I_2/KI solution to the unknown (screen b) to perform one of the tests.

known solutions in order to determine what happens with a ketone or ethyl alcohol. The tests can also be performed on the user's "unknown." Besides the chemical tests, an infrared (IR) spectrum of the compound is available. Finally, students must try to name their unknown, and the computer evaluates their answers.

CHEM DEMO (CDL) (Figure 5.19) contains an acid/base titration, the emission and absorption spectra of various elements, and an IR spectra program. The acid/base program is also on their pH PLOT disk, discussed

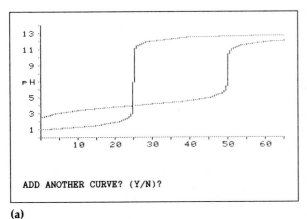

(a)

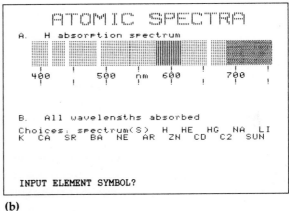

(b)

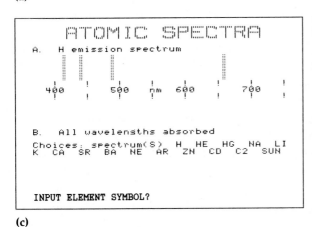

(c)

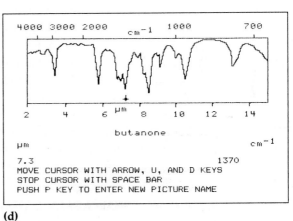

(d)

Figure 5.19 CHEM DEMO contains an acid/base titration program, an atomic spectrum generator, and an IR spectrum generator. Screen (a) shows an acid/base titration experiment; screen (b) shows the hydrogen absorption spectrum from "Atomic Spectra"; screen (c) displays the hydrogen emission spectrum from "Atomic Spectra"; screen (d) shows the IR spectrum of butanone.

below, in a more inquiry-oriented form. The "Atomic Spectra" prism shows either the emission or absorption lines of various chemicals in colorful graphics. The IR spectra program shows the spectra of several organic compounds and allows the learner to move an arrow to get the exact wavelength of the absorption bands.

The Emancipatory Mode

The rapid growth of knowledge in chemistry had made the dissemination of information difficult until the newest technologies became widely available. Public domain and commercial databases concerning the chemical elements, their physical and chemical properties, and chemical reactions are available from a variety of sources, including Project SERAPHIM and TI&IE (Chapter 7). The PERIODIC TABLE (MECC) contains a large database, which helps students quickly locate specific information concerning the elements. HYPERCARD stacks concerning science topics, including the periodic table and chemistry laboratory equipment, are available; and CD-ROMs, which have chemical abstracts and indices, are often available in university libraries.

In traditional chemistry experiments, many students are not actively involved in carefully observing the phenomenon. The use of an MBL frees the student from the monotonous part of the laboratory work and provides a chance to actively investigate. Project SERAPHIM has laboratory modules for making adapter boxes and interfacing devices for measuring temperature, heats of reaction, photochromic kinetics, and colorimetry. HRM, LEAP, Sargent-Welch, Kemtec, and Cambridge Development Lab provide MBLs that allow the teacher to conduct experiments in photovoltaics, high temperatures, Kjeldahl reaction, conductivity, heat of fusion, indicators, pH, titrations, colorimetry, equilibrium constants, and energy and phase changes. CHEMPAC (E & L), a complete course in chemistry, uses the microcomputer as an integral part of the laboratory experiments. These MBLs are discussed in detail in Chapter 8.

The Revelatory Mode

Many companies have produced worthwhile programs on the gas laws. A few programs simply give the name and the formulas. ("This is Boyle's law: $P_1V_1 = P_2V_2$. Now practice"). Several programs, however, provide good graphics that are appropriate for use with the whole class while talking about how the laws were formulated. Individual students can also use these programs to learn about and practice gas law problems.

Prentice-Hall produces two simulations on the gas laws: BOYLE'S AND CHARLES' LAWS and GAS LAWS AND THE MOLE. Help is available, and students can back up one screen to review previous material. The really excellent visuals promote better understanding of the relationships between temperature, pressure, and volume (Figure 5.20). Besides fairly routine drill and practice and instruction, the program presents the ways in

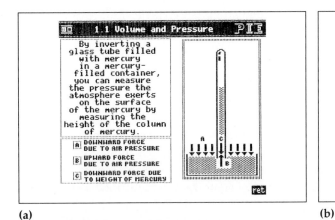

(a)

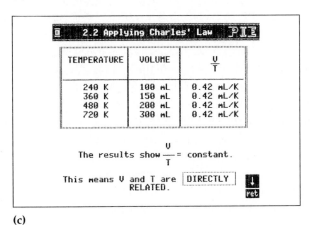

(c)

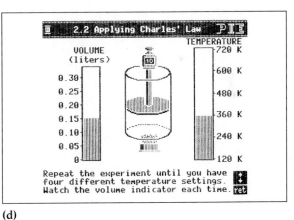

(d)

Figure 5.20 Screen (a), from BOYLE'S AND CHARLES' LAW, demonstrates one of the interactive graphics showing air pressure; screen (b) shows students how Boyle's law was developed; screen (c) shows how Charles' law was developed; screen (d) is an interactive graphical experiment on Charles' law.

which the gas laws were developed, rather than just having students memorize a name and a formula to go with it.

CHEM LAB SIMULATION (High Technology Software) offers programs that include simulated experiments performed by students. CHEM LAB SIMULATION 1 is on titrations. Students perform an acid/base titration, determine the equilibrium constant of a weak acid, or determine Avogadro's number using a titration and a monomolecular experiment. Simulations could be used as a laboratory preparation, prior to actually performing the experiment in the lab or in place of performing one or all experiments. Help with the calculations is available, but the calculations must be performed as if the experiment had been done in the laboratory. pH PLOT (CDL) is a similar program that interfaces easily with MBL technology.

CHEM LAB SIMULATION 2 deals with the ideal gas law and features low-resolution graphics. Colorful molecules bounce around in a container, and students can vary the pressure, volume, temperature, or number of gas molecules. CHEM LAB SIMULATION 3 concerns calorimetry, and the high-resolution graphics are complete with lighted Bunsen burners and mixing reactants. Although the thermometer is difficult to read with great accuracy, it would provide an excellent class demonstration as well as a way of having students perform these standard experiments in class. This package covers heat capacity of the calorimeter, heat of neutralization, heat of solution, and Hess' law.

The final volume, CHEM LAB SIMULATION 4, deals with thermodynamics and covers heat of vaporization and thermodynamics of an equilibrium reaction. Once again, it is difficult to read the bottom of the mercury plug with great accuracy, and the temperature must be controlled very carefully. Nevertheless, it only took a couple of trials for the author to get the computer procedures down to a point where the percent error was a reasonable amount. All four of these disks have a good manual that explains some of the theory and the procedures on the disks.

IT'S A GAS (Diversified Educational Enterprises) is a simulation in which the students must do considerable scientific problem solving. The disk contains an ideal gas law demonstration, some problems on which to practice, a refrigeration lab simulation, and a carbon dioxide fire extinguisher laboratory simulation. The refrigeration lab is a challenge to either the whole class or to individuals. The graphic shows a refrigerator in which someone has placed some food (Figure 5.21). The object is to cool the food as much as possible by varying the amount of Freon in the tubes and in the

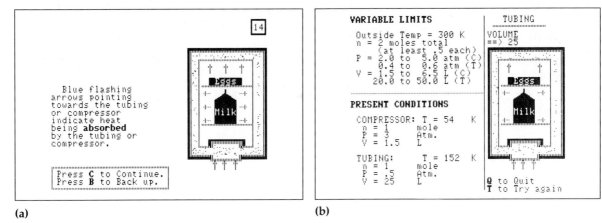

(a)

(b)

Figure 5.21 The refrigeration lab from IT'S A GAS allows students to select the variables for controlling the refrigerator (screen a) and then shows whether heat is being removed from the food in the refrigerator (screen b).

compressor. Users also vary the volume and the pressure in each part of the cooling system. Blue arrows (cold) and red arrows (heat) show whether users have achieved the goal. The carbon dioxide fire extinguisher has a good visual demonstration of the operation and phase changes in an extinguisher; students alter the temperature and the pressure in order to get the extinguisher to dispense the solid foam when the pressure is released.

The Conjectural Mode

CDL's version of the gas laws, EZ-CHEM GAS LAW SIMULATION (Figure 5.22), is the most inquiry-oriented program of all the chemistry programs we reviewed. A "$PV = nRT$ machine" allows free exploration of all variables involved in the gas laws. Students can draw all sorts of graphs on the screen, but they must decide what to see. The program has an ideal gas with a floating plunger and with a controlled plunger, diffusion, diffusion through a grating, the mixing of two gases, and Brownian motion. In the classroom, this is an excellent support to "what if" questions because it will lead to lively discussions without the necessity of calculations. The actual relationships can be derived using the program, but they are derived through an inquiry approach. This program fits well in both the exploration and concept application phases of the learning cycle, with the students and teacher discussing the concepts during the intervening concept introduction phase.

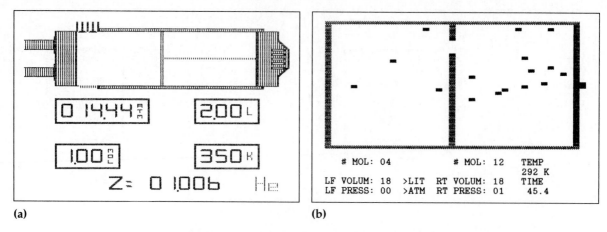

(a)   **(b)**

Figure 5.22 In EZ-CHEM GAS LAW SIMULATION, the $PV = nRT$ machine (screen a), showing the floating plunger, explores the relationships among temperature, pressure, and volume. Screen (b) shows one type of molecule as it diffuses from one side of a chamber through an opening in the barrier.

Physics

If ever there was a course in the secondary curriculum for which the computer was invented, that course is physics. The special capabilities of the computer lend themselves to nearly every aspect of physics. Physics is nearly always the science class in which the fewest students enroll. It seems enigmatic that physics historically led the way into the Renaissance—the scientific revolution. These ideas, "laws," and formulas, which our current students find so abstract, were those that the great scientists such as Galileo, Newton, Kepler, and others found so concrete, challenging, and useful.

The computer can pull together the massive amounts of data generated by physics experiments. The computer can show graphically what different equations mean and what they describe under different conditions. The computer can draw diagrams and simulate real-world information, which students often simply try to memorize as formulas.

The Instructional Mode

Many of the programs designed for physics drill and practice or tutorial work are written for college physics classes. As a result, teachers need to review the programs in order to make sure that the level of difficulty is adequate and that the programs are reasonably self-explanatory. Newer programs are more user-friendly and often contain some graphics that help explain phenomena.

GENERAL PHYSICS, a set of twelve volumes (Cross Educational Software), covers many physics topics (Figure 5.23). These include vectors, statics, motion, conservation laws, circular motion, thermodynamics, electricity, optics, atomic physics, solar system and stellar astronomy, and physics gems. These programs present algorithms for solving problems in a step-by-step format with examples. The explanations of what to do could help those students whose strategy consists of memorizing the formulas. Problems are included with all disks, and all phenomena are highly visual. Animation, color, and some games help keep the disks interesting. Short instruction sections are included for all topics, and short manuals are available for each of the twelve disks.

The Emancipatory Mode

The AIR TRACK SIMULATOR (CDL) presents a graphical view of one-dimensional collisions. It is a good tool whether or not there is an air track in the classrrom. It offers several demonstrations, allowing students to enter their own parameters and observe the resulting collisions. The tutorial section explains how to determine velocity, momentum, kinetic energy, potential energy, and the like. Students can even design their own collisions from a parameter list.

Chelsea Science Simulations has produced a program called SATELLITE ORBITS, which is distributed by Conduit. The object of the program is to calculate the path of a satellite injected into orbit horizontally. Students can vary the velocity and the height of the injection and are challenged to produce a circular orbit around the Earth's surface. The program can have

Figure 5.23 Screen from MOTION, volume 3 of GENERAL PHYSICS.

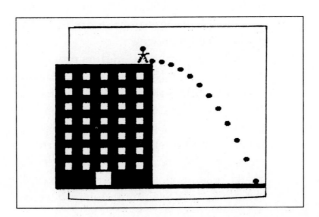

students apply calculations on motion in orbit already learned or have them use trial and error to solve orbital motion problems, thus freeing students from the calculations and promoting an intuitive understanding of satellite orbits.

MBL packages, discussed in Chapter 8, also provide an option for physics students. Several experiments that are tedious or difficult are greatly enhanced with the use of devices interfaced to the computer. The American Association of Physics Teachers has produced a manual for teachers, and their journal (*The Physics Teacher*) has often included articles on the construction and use of MBLs. The MBL package MOTION (HRM) provides an efficient and accurate method of studying motion without lengthy data collection and manipulation to arrive at possible formulas that describe the motion. Vernier, Cross Educational Software, Thornton Associates, and LEAP also produce several different physics MBL packages.

Programs that graph data collected and entered into the computer during an experiment are described in Chapter 7. These packages are useful in helping students determine which formula best represents the data collected in an experiment. In addition, using the computer to create graphs allows students more time to reflect on the relationships identified on the graph instead of spending their time on the construction of the graph itself.

The Revelatory Mode

INTERACTIVE OPTICS (EduTech) introduces geometric optic principles using colorful simulations and interactive demonstrations. Each section begins with an inquiry mode, where students investigate the particular phenomenon for as long as they wish before proceeding to the tutorial. Four major programs cover thin lenses, reflections from a plane mirror, Fermat's principle, and Snell's law. The high-resolution graphics would work equally well in a whole-class discussion or for individual students just starting the topic or needing review.

After the lenses and mirrors programs, a motivating and challenging game of "Hide and Seek" is available. During this game the computer hides somewhere on a grid, and the learner tries to find the computer by moving around on the grid to find out if the computer can see him or her. To aid in the hunt, there is a numbered grid on which the computer crosses out the spots that absolutely cannot be the correct answer on each turn, based on what the computer can "see." Fermat's principle is also tested through a gamelike simulation. Instead of simply dealing with refraction when a light beam leaves one medium and enters a second, the user is shown a drowning swimmer and must determine how to get to him as fast as possible. Because better time is made when running on the beach than when swimming, the learner can choose to enter the water at various points. The computer calculates the time it would take to run along the beach plus the time swimming in the water to reach the drowning person the fastest. The simulation shows the path the runner/swimmer takes and relates that to "refraction," which occurs as the runner goes down the beach and then enters the water (a different medium).

Vernier produces seven different physics labs, each with its own teacher's guide. VECTOR ADDITION III adds or subtracts vectors. After students enter magnitude and direction for each vector, the program displays arrows representing the vectors and the resultant. In KINEMATICS II, the student controls the motion of a truck as it travels across the screen (Figure 5.24a). The student is given some information and tries to produce the desired motion. When the trip is completed, the results are analyzed, and the student is given some feedback on the results.

PROJECTILES II challenges students to learn about projectile motion by hitting a target at a known range by launching off a 100-m cliff (Figure 5.24b). ORBIT II has nine challenges that allow students to experiment with satellite motion. Students obtain a circular orbit by launching from a space station or use thruster rockets to correct the elliptical orbit of a satellite.

CHARGED PARTICLES II allows students to experiment with the motion of an electrically charged particle in magnetic and electric fields. Using several different particles, students experiment with their field strengths and speeds. One might find the mass of an electron by experimenting with how it is affected by a magnetic field.

RAY TRACER draws ray diagrams to illustrate the principles of geometrical optics (Figure 5.24c). It can be used as a demonstration or as an educational game. It demonstrates Snell's law, reflection, dispersion, mirrors, lenses, images, and spherical and chromatic aberration.

WAVE ADDITION III graphically demonstrates the superposition of waves (Figure 5.24d). The program operates in nine different modes, some of which are programmed to demonstrate various principles. In other modes, students select the waves to be added. Component waves and the result of the superposition are plotted on the screen. The program demonstrates the relationship between frequency, wavelength, and wave speed; constructive and destructive interference; beats; Fourier synthesis of a sawtooth wave, square wave; and so on.

High Technology Software produces PROJECTILE MOTION WORK-

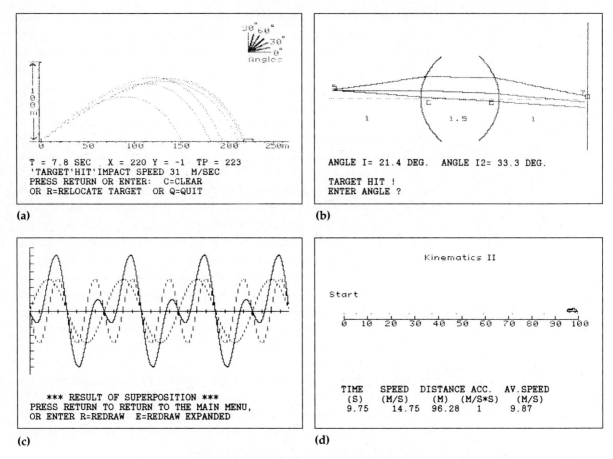

Figure 5.24 Screens from several Vernier programs. Screen (a) shows the acceleration of a truck in KINEMATICS II; screen (b) shows the attempt to hit a target in PROJECTILES II; screen (c) shows the refraction of light through a thick lens in RAY TRACER; screen (d) is a demonstration of the addition of two waves in WAVE ADDITION II.

SHOP, STANDING WAVE WORKSHOP, HARMONIC MOTION WORK-SHOP, and CHARGED PARTICLE WORKSHOP. Like the company's chemistry lab simulations, these programs make excellent visuals for whole-class display while discussing these phenomena. In addition, the programs can be used in an inquiry mode, if desired. Students can vary parameters to observe the various motions and can try to verbally describe the relationships.

PROJECTILE MOTION WORKSHOP is a collection of four programs designed to illustrate motion under the influence of a uniform force of gravity, ignoring friction. Vertical motion, with the object dropped from rest or fired both in an upward direction and downward direction, can be shown in a graph with a velocity vector or a graph of the kinetic and potential energy indicated beside the position graph. Fired upward and downward programs show two-dimensional motion, where students can vary the angle above or below the horizontal and the initial velocity. A velocity vector can be shown as well as the kinetic and potential energy graphs. The final part of this program is component motion, which demonstrates that the motion of a projectile is a combination of motions.

The STANDING WAVE WORKSHOP exhibits the motion associated with a vibrating string that is fastened at both ends. The program graphically displays the vibrational characteristics of ten harmonics, individually as well as combined. The program allows users to select the point at which the string will be plucked and shows, in graphical form, the harmonics produced, while the high-resolution outline of the wave shows the motion of the string. In HARMONIC MOTION WORKSHOP, simple and damped harmonic motion are studied. After placing an object in harmonic motion on the screen, students can alter phase, amplitude, and the damping factor. Students can draw instantaneous velocity and acceleration vectors while the object is in motion. The program also displays the behavior of the object if it were in circular motion.

The CHARGED PARTICLE WORKSHOP is a series of three programs that simulate the motion of charged particles under the influence of various combinations of electric and magnetic fields. The display shows the x- and y-components of velocity and acceleration. The fields, the initial velocity, and the mass of the particle and its electric charge can all be varied as the motion is studied.

The Conjectural Mode

CIRCUIT LAB, written and distributed by a physics teacher (Mark Davids), is an interesting way of approaching the learning of direct current circuits (Figure 5.25). The program allows students to build and test their own circuits in a hands-on/minds-on manner. The program can be used by the whole class, but it is designed for individual or small-group use. Three basic circuit layouts are available—series, parallel, and combinations. Once students have selected a general layout for the diagram, they place resistors, light bulbs, voltmeters, ammeters, and switches in the circuit. Students

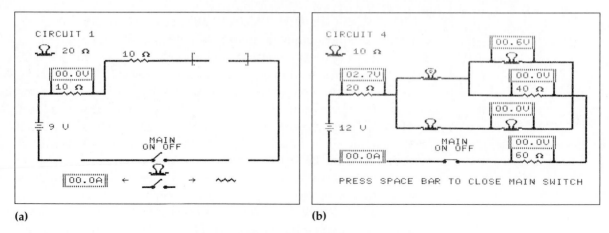

(a)   **(b)**

Figure 5.25 Screens from CIRCUIT LAB. Screen (a) shows the design of a simple circuit, and screen (b) shows the testing of a complex circuit.

select the values of the resistors and the voltage of the battery that will power the circuit. After students construct and activate their circuits, lights come on, and ammeters and voltmeters show the values of the current or the voltage at that point of the circuit. Students then go back and modify and repeat the process as they build up intuitive ideas about direct current.

The three levels of complexities in the program provide increasingly more difficult analyses. Once students have developed ideas about series and parallel circuits and the voltage and current going through them, they are ready to discuss what they think the mathematical relationships are. When the relationships are known, a circuit can be built and the currents and voltages predicted. The switches can then be thrown, and the values checked on the appropriate meters.

A-MACHINE and INCLINED PLANE, two programs in the LAWS OF MOTION series distributed by EME and developed at the University of Pittsburgh, help students study motion (Figure 5.26). A-MACHINE is a study of the motion of a block along a plane surface in two different worlds. These two worlds have different laws of motion. The block sits on a plane surface attached to a bucket of sand by a string. The bucket of sand hangs off the plane surface so that it can fall when released. In addition, the string to the bucket can be cut at any point along the plane surface.

The program, designed for individual use, allows students to develop intuitive ideas about the laws of motion. The surface of the plane may be

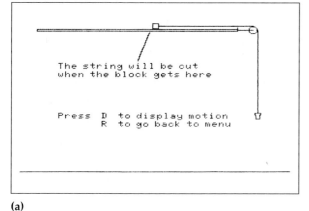

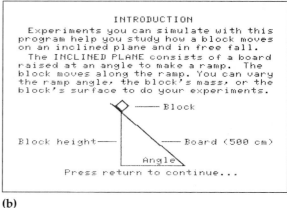

(a)   (b)

Figure 5.26 In screen (a), the bucket of sand pulls the block along a plane surface in A-MACHINE. Screen (b) shows the introduction to INCLINED PLANE and explains the variables the student can manipulate.

rough or smooth. Students can decide on the mass of the sand and the point at which the string will be cut. In one world, the mass of the sand in the bucket determines the final velocity of the block and the distance traveled—provided, of course, that friction was overcome (Figure 5.27). In the other world, objects have a constant velocity under many conditions.

In INCLINED PLANE, students study free fall and motion along an inclined plane under the same two conditions described in A-MACHINE. One world appears to resemble the one in which we live, and the other appears to have uniform velocity both under free fall and when sliding down an inclined plane. Forces of friction and the mass of the object are varied in both worlds to see how these affect the movement of the blocks.

INTEGRATION OF SCIENCE WITH OTHER SUBJECT AREAS

As we indicated in the previous two chapters, there are many opportunities to integrate science with reading, language arts, mathematics, and social studies. Critical thinking skills emphasized in science software can be applied to these other content areas. In addition, when students write their own reports and read other students' laboratory reports, many important reading and writing skills are developed and practiced. The language used when students talk with each other as they experiment helps them practice

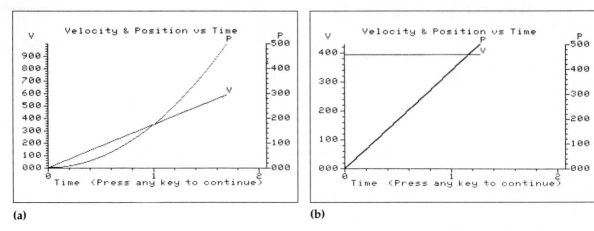

(a)   **(b)**

Figure 5.27 Screens (a and b) from INCLINED PLANE show identical setups of the inclined-plane experiment. In both cases, the block mass is 500 g, the angle of the ramp is 45 degrees, the height of the block is 353.6 cm, and the plane surface is rough. The graph in screen (a) represents the block sliding down the ramp in one microworld; screen (b) represents the other microworld.

clear, logical, and well-organized communication, contributing to important language arts learning. The opportunities to increase functional vocabulary when conducting science experiments are great. Science experiences provide an excellent "jumping off place" for language arts, reading, and mathematics lessons. The computer enhances all these connections with the use of simulation, problem solving, database programs, and telecommunications.

Because of departmental structures and specializations of teachers, this integration is sometimes more difficult to accomplish at the secondary level than at lower levels of education. Educators interested in promoting the integration of science with other secondary subject areas—for ideas on how reading and language skills can contribute to the science curriculum—may be interested in many of the programs and strategies described by Schwartz and Vockell (1989), and King and Vockell (1991). Similarly, many of the programs discussed by Dinkheller and associates (1989) can support the integration of mathematics and science skills. Brown and Vockell (1990) include discussions of social studies programs that can easily be integrated with the science curriculum—for example, geography programs can often overlap with Earth science objectives. Finally, Vockell and van Deusen (1989) describe numerous specific programs and strategies for transferring higher-order thinking skills across various curriculum areas.

SUMMARY

This chapter has presented a sample of the many programs that are available to help enhance secondary science instruction. We have described many of the programs in detail, to convey an impression of what is possible. These programs can provide powerful tools for teaching traditional science topics. To employ them as usefully as possible, teachers should apply them in conjunction with the principles discussed in Chapters 2 and 9.

REFERENCES

Aldridge, B. G. "Essential Changes in Secondary Science: Scope, Sequence, and Coordination." *NSTA Report* 1 (January/February 1989): 4–5.

Brown, W., and E. L. Vockell. *The Computer in the Social Studies Curriculum.* Watsonville, Calif.: Mitchell, 1990.

Dinkheller, A., J. Gaffney, and E. L. Vockell. *The Computer in the Mathematics Curriculum.* Watsonville, Calif.: Mitchell, 1989.

Herron, J. D. "Piaget for Chemists." *Journal of Chemical Education* 52 (1975): 146-150.

King, R., and E. L. Vockell. *The Computer in the Language Arts Curriculum.* Watsonville, Calif.: Mitchell, 1991.

Rivers, R., and E. L. Vockell. "Computer Simulations to Stimulate Scientific Problem Solving." *Journal of Research in Science Teaching* 24 (1987): 403-415.

Schwartz, E., and E. L. Vockell. *The Computer in the English Curriculum.* Watsonville, Calif.: Mitchell, 1989.

Vockell, E. L., and R. van Deusen. *The Computer and Higher-Order Thinking Skills.* Watsonville, Calif.: Mitchell, 1989.

Wise, K. C., and J. R. Okey. "A Meta-Analysis of the Effects of Various Science Teaching Strategies on Achievement." *Journal of Research in Science Teaching* 20 (1983): 419-435.

CHAPTER 6

PROBLEM SOLVING AND THE SCIENTIFIC METHOD

THE TOPIC OF TEACHING higher-order thinking skills is currently at the center of attention at all educational levels throughout the United States. Another book in this series is devoted entirely to this topic (Vockell and van Deusen, 1989). Most of the skills typically discussed as "higher-order thinking skills" are incorporated as necessary steps in scientific problem solving. Many of the programs described in the preceding three chapters help students develop and practice these skills. This chapter focuses specifically on the thinking skill aspects of science programs.

SCIENTIFIC THINKING

Many students have a limited view of the scientific enterprise. They believe that scientists use special equipment and do experiments that lead directly to useful inventions and cures (Educational Technology Center, 1988). Students often have no notion of science as the intellectual construction of theories, nor of scientists engaging in public discussion on points of disagreement in their hypotheses. These mistaken notions are supported by the textbooks students use, which often present science as a "rhetoric of conclusions."

As young students try to solve problems in science, they often exhibit much undisciplined trial and error. They may simply make wild guesses to solve problems, without even keeping track of what they just tried; or when faced with continued faulty solutions, they may continue to change one or more variables at a time without any way of knowing whether they are getting closer to the correct answer. Many students simply feed variables into a formula with no real sense that the problem is organized around key interpretive concepts. They spend their time memorizing facts and rules about when this or that applies. This is not the same process used by scientists who use "trial and learn" (as the "Search for Solutions" film series calls it) when they are in a problem-solving framework.

Most science courses would list as one objective that students be able to perform an experiment showing an understanding of the scientific method and scientific problem solving. This means that students should be able to define the problem, suggest several alternative solutions, plan a problem solution or select procedures, implement the solution, analyze the results to see if the desired result has been obtained, reformulate the hypothesis and the procedure, if necessary, and repeat the experiment using the new procedure. This process should continue until a satisfactory conclusion has been obtained.

Computer-learning environments are ideally suited to imparting general problem-solving strategies and higher-order thinking skills because they can provide models, based on sound scientific theory, to stimulate thinking and discussion about the thought processes involved in solving problems. Defects of poor strategies can be explained to students, and suggestions can be made if students cannot think of alternative paths to try. Many of the programs and computerized strategies discussed throughout this book teach scientific thinking skills. This chapter presents those programs that focus specifically on higher-order thinking skills. In many cases, it makes sense to teach these skills as separate instructional units and then to focus on them as playing an important role in scientific processes.

Instruction in the scientific method must provide opportunity and motivation. It must give students an opportunity and urge to develop and employ process skills rather than focusing purely on the content emphasized in many classrooms. Promoting this scientific mindset is difficult when the problems that students are asked to solve are usually "school problems," in which students are rewarded for applying an almost ritualistic formula. Stereotypical problem solving and memorization do little to help students generalize their problem-solving notions to everyday nonschool problems.

Learning to engage in scientific thinking requires the mastery of process skills rather than content skills. As the name indicates, process skills focus on general mental processes (e.g., how to generate new solutions) rather than on the informational content covered by a given lesson. Another way to state the distinction is that process skills involve procedural knowledge, whereas content skills involve declarative knowledge. Students engaging in process skills must often think divergently in order to "invent" or "discover" concepts. Content skills, on the other hand, often require students to think convergently because there is a single, right answer. Teachers should use process strategies (including process-oriented computer programs) to teach the scientific method, a general concept, or a thinking skill in isolation. Later, the scientific method, the concept, or the skill can be applied to particular topics.

To develop effective problem-solving strategies, students need practice in the processes involved in solving problems. Teachers must model the process of problem solving, not just ask students to do homework. In addition, students need multiple different examples to provide a groundwork for generalization or transfer of problem-solving abilities.

Experts generally agree that an inquiry or "guided discovery" approach is the best strategy for teaching thinking skills (e.g., Mallan and Hersch, 1972; Rosenshine and Furst, 1971). When this approach is used, one of the best methods for keeping students involved in the process is to respond nonevaluatively. This means the refusal to label their responses as good or bad, as right or wrong, or as interesting or dull. It is a good practice for teachers to use a nonevaluative response strategy when teaching process skills even without the computer, but the computer sometimes makes this kind of strategy easier to implement. The teacher can let the *computer* do the evaluating. If a student is wrong, let the computer show the natural consequences. If the student is right, let the computer show this as well.

This nonevaluative strategy may fly in the face of what many of us have been taught: namely, to praise student responses consistently and to find good points in all their responses. This extreme positive attitude may be appropriate in content learning; however, it is often counterproductive in process situations. In process learning, it is important to establish a risk-free environment in which students can hypothesize and test, infer and predict. Psychological safety and freedom are necessary components of a classroom climate capable of fostering higher-order thinking skills (Rogers, 1961). If the teacher compliments a student for a particular type of thought or response, we can be sure that the teacher will receive many more responses in that same pattern of thought. Because the goal is internalization of procedures or divergent thought and creative thinking, it is usually inappropriate to channel responses into narrow paths or to promote simple duplication of strategies tried by other students. Students *will* be rewarded for successful thinking, and other students will recognize this; but the reinforcers will be the intrinsic rewards that come from successfully accomplishing a difficult task.

Finally, when teachers are effective at teaching problem solving, it is usually because they incorporate a strategy for transfer of learning. They usually employ the "remember when" strategy discussed later in this chapter or at least specifically direct the students' attention to the activities that have enabled them to be successful in an activity. For example, in a microcomputer-based laboratory (MBL) project with junior high students, the teacher might require students to make a prediction about the results of the next experiment based on their previous experiences. These students would be likely to have a more robust understanding of factors involved in the new unit than those who merely observed carefully (Friedler et al., 1987).

THE SCIENCE PROCESS SKILLS

Scientific thinking is not a single skill, but rather the orchestration of several separate process skills. These skills include observing, classifying, communicating, drawing inferences, and generalizing. The rest of this chapter describes several computer programs that can stimulate such skills.

GERTRUDE'S SECRETS and GERTRUDE'S PUZZLES (The Learning Company) both deal with the early elementary skill of classification and attribute identification. Not only must students recognize which attributes the various objects have, but they must also recognize how these attributes are similar and different in order to solve the puzzles they are given. GERTRUDE'S PUZZLES is like a house with many rooms (Figure 6.1). Each room holds a different puzzle to solve. Each puzzle is more complex than the one before; and as they proceed through the puzzle, students must learn increasingly complex ways of using the characteristics of the items. The program requires discriminating shapes and colors, identifying relationships between objects, categorizing, analyzing patterns and rules, using deductive reasoning, and discovering multiple solutions to problems. GERTRUDE'S SECRETS builds on these skills to form more complex puzzles requiring better discrimination of similarities and differences and in discovering the rules that govern the puzzles.

THE INCREDIBLE LABORATORY (Sunburst) puts students into an environment in which they can develop and use problem-solving skills. While some programs develop memory and discrimination, attributes, and rules, THE INCREDIBLE LABORATORY (Figure 6.2) is designed to help students learn to work with hypothesis-testing strategies. Students must scan data in order to gather more ideas and/or eliminate possibilities. They must make organized notes in order to remember what they have tried. They must analyze pictures for pattern or sequences, seeking information that will lead them to a solution.

Students using THE INCREDIBLE LABORATORY use "trial and learn" to discover how eighteen "chemicals" contribute to the creation of unusual monsters. Each chemical determines the variation of one of the monster's six key body parts—the head, eyes, arms, body, legs, or feet. Through three successively difficult levels, students build monsters and then try to determine what chemicals caused the various features. In the most difficult mode, the "Scientist," the effects are different each time the program is run, so that students cannot rely on what they found out the day before. In addition, in the "Apprentice" and the "Scientist" modes, students use combinations of chemicals that interact with one another and try to determine how the monster is affected.

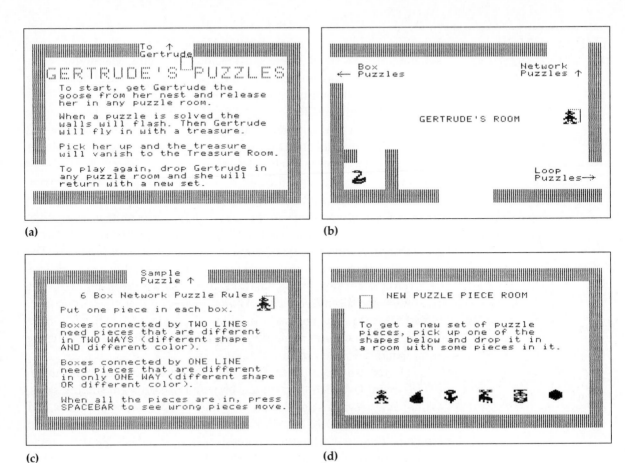

Figure 6.1 GERTRUDE'S PUZZLES (screen a) is an elementary program that presents attribute puzzles located in different rooms (screen b). The puzzles in each room become increasingly more complex (screen c). The student can redesign some of the problems in the "New Puzzle Piece Room" (screen d).

THE INCREDIBLE LABORATORY is an example of a good program that is often a complete waste of time unless it is introduced to students properly. We have seen teachers who have told students to "go run THE INCREDIBLE LABORATORY" as a reward for finishing worksheets early. Under these circumstances, students almost inevitably think the monsters are cute or interesting at first, then they guess wildly or randomly to try to make new monsters, but they almost never get around to employing a true scientific method to solve the problem as described in the "Monsters" box

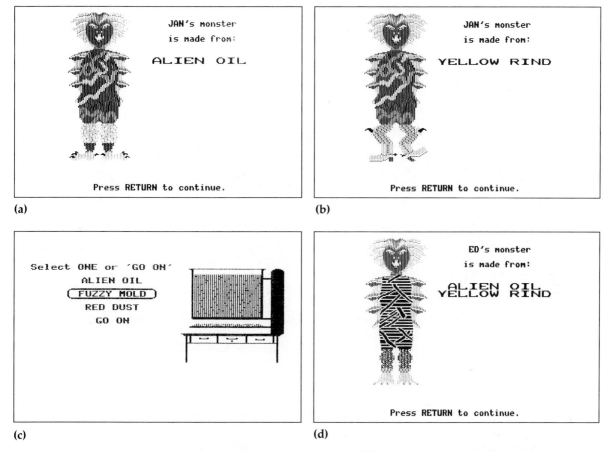

Figure 6.2 Screens (a and b) from THE INCREDIBLE LABORATORY show the result of choosing only one chemical to create the monster. Screen (c) shows the fifteen different chemicals in sets of three from which the student selects. Screen (d) shows the monster created from two chemicals.

on pages 156–157. The sensible way to use THE INCREDIBLE LABORA-TORY is to employ it as part of a planned unit of instruction. The program could be used during the learning stage (exploration and term introduction) of instruction, with the teacher using a guided discovery approach to help students focus on the value of varying only one factor while holding all others constant. The program could also be used during the practice phase (concept application) of instruction, with the students practicing this aspect of the scientific method after it has been introduced in some other context.

Monsters

The students in Miss Steinberg's class were using THE INCREDIBLE LABORATORY to create monsters. The students named chemicals for her to feed into the computer to create monsters, which were projected on a large screen in front of the class. She ran the program twice, and the students were delighted with the "gross" monsters they were able to create. Miss Steinberg then placed $1000 in play money on her desk and announced that she was willing to bet that she could beat them at the challenge: the participants entered chemicals, the computer generated the monster, and they had to guess which of three monsters matched the chemicals fed into the computer.

One student asked if she could conduct some experiments first, and Miss Steinberg acquiesced. The conversation went like this:

MARIAN: The monster on the screen was made from black ice and sparkles. Let's change a few of the chemicals and see what we get.

JUAN: Let's take notes about this monster first. Let's

see. This one has a head with no ears, a twisted smile, and . . .

MEG: Just call it a skeleton head. We don't need all the details.

The students then made up names describing the monster's head, torso, arms, legs, and feet. Then they made a new monster with goose grease and alien oil. The resulting monster looked nothing like the first. They made up new names to label what they saw. They wrote down the chemicals and listed the output.

KATIE: This is getting us nowhere.

COURTNEY: That's because we're changing too many things at the same time. Let's do everything the same as last time, but put in yellow rind instead of goose grease.

KATIE: That's a good idea. But here's an even better one. First, let's make a Super Monster with *all* the chemicals. Then we'll make five more monsters, each time leaving out one of the chemicals. When

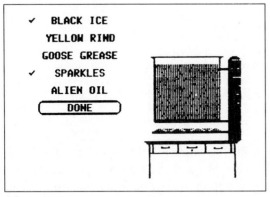

(a)

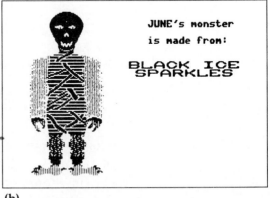

(b)

Screen (a) shows THE INCREDIBLE LABORATORY chemical selection screen. Students select chemicals from this screen to create a monster such as the one shown in screen (b). Screen (b) shows a monster made with black ice (brown leg-warmers) and sparkles (blue, four-toed feet).

(Continued)

we leave out a chemical from the Super Monster, we'll know that chemical caused whatever has changed.

The students tried Katie's method, and within five minutes they had bankrupted the teacher. Miss Steinberg pointed out that giving names to the body parts had really made things easier. She said that the students had developed a sort of scientific terminology for monsters. She also pointed out that the method they had used was very similar to the process that real scientists had used to discover a vaccination for smallpox in the chapter they had read two weeks earlier.

THE FACTORY (Sunburst) (discussed in detail in Vockell and van Deusen, 1989) asks students to produce a particular product in their simulated factory (Figure 6.3). To do this, they must use "machines" in a proper order to apply a particular pattern of painted stripes, turn the product, punch holes, and perhaps paint again. Students can see their final product and compare it to the one they were asked to produce. If they were not successful, they can try again. Depending on how this program is used, students can learn or practice skills of hypothesis testing, sequencing, backward planning, generating multiple solutions, and using a physical model as part of the planning process. These are all valuable skills in both scientific experimentation and everyday problem solving.

THE FACTORY and THE INCREDIBLE LABORATORY are accompanied by large, carefully designed notebooks, which include teacher's guides and student worksheets. In addition, these and other Sunburst programs are also accompanied by a large chart called the "Problem Solving Matrix" (Figure 6.4). This chart shows the teacher which skills the program is designed to foster. Individual programs are accompanied by more detailed versions of this matrix, showing exactly what skills each program is designed to teach. Our experience is that these individual program charts are extremely useful but slightly inaccurate. In general, all skills indicated on a chart are adequately covered by a designated program, but often additional skills are covered as well. In addition, a teacher who wishes to promote a higher-order thinking skill can often accomplish this by making slight adjustments in the lesson plan that actually implements the program.

Several Sunburst programs have videotapes, which the teacher can obtain to view methods of using the software in the classroom. As with any classroom resource, the careful integration of the computer software into

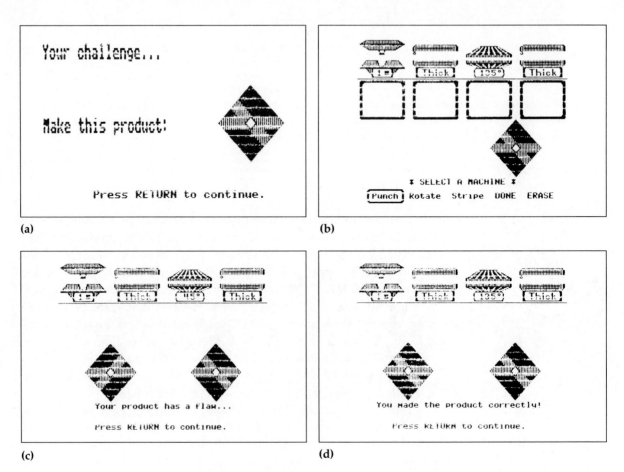

Figure 6.3 Four screens from THE FACTORY. The student is instructed to manufacture the product shown in screen (a) by using a series of machines, as shown in screen (b). A result of using the machines in screen (b) incorrectly would be the flawed product, shown in screen (c). The correct solution is shown in screen (d).

the total learning cycle helps students learn. Computer software without any critical planning by the teacher is only a diversion, at best.

Teachers will be more effective at teaching higher-order thinking skills if they are familiar with the model shown in Figure 6.4 or with the principles described in this book and in Vockell and van Deusen (1989). For thinking-skill programs to be effective, it is crucial for teachers to know what skills

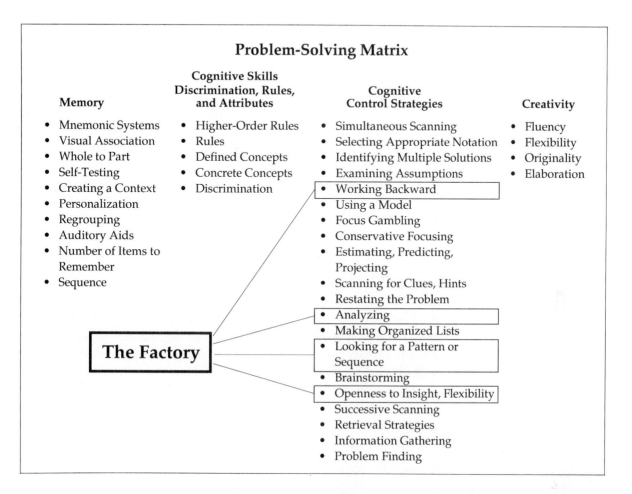

Figure 6.4 This chart shows the thinking skills typically covered by THE FACTORY.

a program can help students learn, to help students focus attention on these skills while running the programs, to help students supplement the programs and integrate them with previous learning as necessary, and to provide opportunities for students to generalize the skills to different situations.

INTEGRATING THE SCIENTIFIC METHOD

Although it is possible, and often desirable, to teach the scientific method using programs that are relatively content free, it is also important to incorporate the scientific method into units focusing on the scientific subject matter of a course. One computer program that begins to help students use the scientific method is DISCOVERY LAB (MECC) (Figure 6.5). This program introduces the student to observation, experimental design, and hypothesis testing. As the program begins, a spaceship arrives from deep space with some alien organisms unknown to Earth scientists. The objective for the students is to study the organisms by altering environmental conditions in a test chamber.

The program has three levels of play—"Training Lab," "Explorer Lab," and "Challenge Lab." Each lab contains five organisms with which to experiment. In the "Training Lab," there are only two variables—light and temperature. In the "Explorer Lab," there are five variables—light, temperature, sound, food, and moisture. In the "Challenge Lab," the variables are the same as in the "Explorer Lab"; however, each variable has three different settings. These variations in level of complexity permit the program to be used at different grade levels, or they permit the instructor to provide simpler programs for students who have not yet mastered skills (remedial activities) or more advanced problems for students who master skills quickly (enrichment activities).

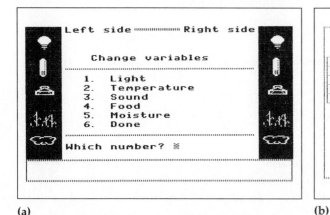

(a)

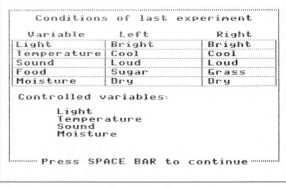

(b)

Figure 6.5 Screen (a) shows the "Explorer Lab" setup menu in DISCOVERY LAB. Screen (b) shows an analysis of all variables that have been set in the "Explorer Lab."

Students—or better yet, small groups of students or the whole class with teacher guidance—select the level of difficulty and the organism with which to work. They introduce a number of these organisms into the chamber and vary the conditions. For example, in the "Training Lab," students might turn on the light on one side of the chamber and off on the other side. They might have the temperature cool on both sides of the chamber. After some length of time, students observe where the organisms have congregated, make some notes, and decide if the organisms prefer the light or the dark. If neither the light nor the temperature variable is changed, the computer reminds students that in a scientific experiment at least one variable should be changed at a time (Figure 6.6). If they do not use enough organisms or if they do not wait long enough, the computer reminds them of good scientific procedures. When all variables in the chamber have been tested, students can report their results to the scientific community. After entering their conclusions, they are told if other scientists agree with them.

DISCOVERY LAB challenges students to determine the observable characteristics of their organisms within an inquiry frame of reference. As students experiment, they learn about the processes involved in scientific experimentation. If they proceed through all three levels of difficulty, students must focus on control of variables, making careful observations, keeping accurate records, and drawing conclusions from sometimes contradictory data.

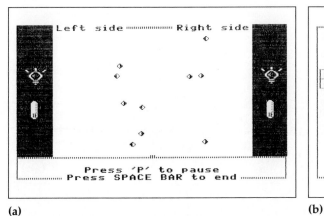

(a)

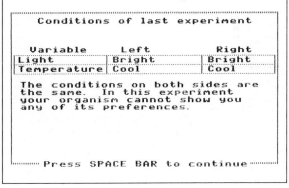

(b)

Figure 6.6 Screen (a) shows ten organisms in the "Training Lab" in DISCOVERY LAB. Both sides of the chamber have bright light and low temperature. Screen (b) shows an explanation of the error in setting the chamber variables.

Each level of the program is unique, and no organism reacts the same at all levels. In addition, all conditions other than those used in the program are controlled. Identical experiments can produce varied results, indicating the need for multiple trials of the same experiment. Because organisms vary in behavior, final conclusions cannot always be determined from just one trial.

This program can be used in the same way any long-term science project might be carried out in the classroom. Students need to keep journals or lab notebooks, and the teacher needs to assume the role of a facilitator and not give students correct answers. Student handouts and teacher explanations are provided to help with the implementation of the program in the classroom. An additional challenge can be added to the "Explorer Lab" and "Challenge Lab" levels by asking students to rank order the five variables being tested.

DISCOVER: A SCIENCE EXPERIMENT (Sunburst) is similar to MECC's DISCOVERY LAB. This program (Figure 6.7) imagines that the World Space Federation probe has just returned with a number of alien organisms frozen in cryogenic pods. Some of these creatures are thawed and allowed to interact in a highly controlled chamber. The challenge to the student is to discover what conditions are necessary for these life forms to survive. This program contains more animated graphics than its predecessor from MECC. The laboratory into which the creatures can be placed contains a

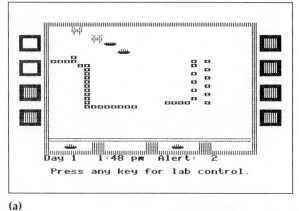

(a)

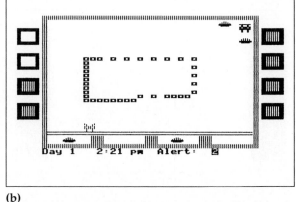

(b)

Figure 6.7 Screen (a) shows two creatures in the laboratory in DISCOVER: A SCIENCE EXPERIMENT, and one is eating. In screen (b), the creature in the upper right is "sick."

robotic arm to open various types of food for the creatures. The arm can also raise and lower barriers built into the lab's floor in order to isolate various creatures from each other.

While there is no time limit on the experimenting, some organisms don't last very long in the chamber without help. Sick creatures are automatically returned to their pods where they recover. Students are allowed to add more creatures at intervals; and if they continue long enough, they may try to work with eight creatures all at once. To do this, students must have patience, good powers of observation, good understanding of variables, and the ability to analyze the information they receive as a result of the creatures' behaviors.

Data diskettes can be prepared and games can be saved in order to work for longer periods of time. Students may become frustrated; thus, the teacher may wish to have class discussions in order to share what techniques might be tried. Teachers can challenge students' ideas, asking them what proof led them to their conclusions.

The teacher's manual contains several good ideas for the teacher to use. Pertinent questions are included to help guide students without giving them an answer. For example, "Do the creatures keep up a cycle of hunger or other behavior that seems to depend on the clock?" The teacher also has a guide to the creature personalities, although it cautions about providing these clues. An example of the creature personality is "Creature 3 is a pest. It will chase Creature 1 and hound it to death."

VISUAL ILLUSIONS (HRM) (Figure 6.8) is a program that also helps students learn the fundamental tools of scientific inquiry. This program differs markedly from the "Discovery" programs in that students actually design the whole experiment and do not work with one that has already been designed. This program lets students display and manipulate six different visual illusions to measure the strength of each, by developing hypotheses and designing experiments to test them.

No single measurement can help them determine the strength or persistence of a particular illusion. Students collect data from a series of trials with various individuals. They then must learn to analyze scattered data from these multiple measurements. Students can create histograms from data that have been saved on a data disk, calculate averages and percent differences, and merge two sets of data into a single file. Both the data and the figures can be printed. A screen editor allows students to change the illusions themselves to try to discover what actually causes the illusion.

One illusion with which students work is the "T Illusion" in which the length of the stem of the T is drawn so that it looks like it is the same length as the cross member of the T. If the two lines are drawn so they appear to be the same length, they will measure different lengths. If the T is drawn so that the lines measure the same length, they will not appear to be the same. Other illusions are the "Muller Lyer," "Parallelogram," "Perspective," "Poggendorg," and "Garden Path." A good teacher's manual with student worksheets is included along with explanatory material. The

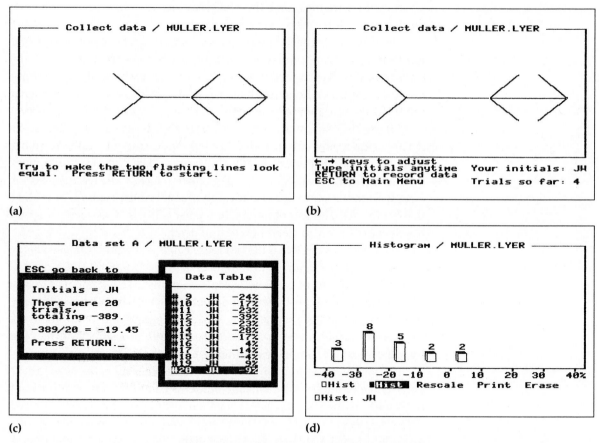

(a)   (b)   (c)   (d)

Figure 6.8 In screen (a) from VISUAL ILLUSIONS, the student selects the "Muller.Lyer." Screen (b) shows the student how to use the arrow keys to change the length of one of the horizontal lines. Screen (c) presents a data table following an experiment on the "Muller.Lyer," showing the average student error. Screen (d) provides a histogram of the same experiment.

program follows all phases of the learning cycle, from exploration through application.

STAR SEARCH (Earthware Computer Services) (Figure 6.9) involves the exploration of newly discovered planets around the fictional star Epsilon Eridani. The program, designed for middle school and up, begins with the selection of a crew and provisions for the trip. The objective is to discover some forms of extraterrestrial life and return to home base safely.

Students can perform various scientific experiments at each planet that is visited. Teams of students can do an atmospheric analysis, a radio telescope scan, infrared/ultraviolet scan, visual surveys with a telescope, gravimeter studies, and magnetometer studies from an atmospheric probe. Part of the crew can land on a planet and do a visual survey, perform gravimeter and magnetometer studies, drill into the subsurface, take pictures with a camera, analyze the surface material, and set up a seismometer. The crew is selected for its ability to interpret these tests, but the students must have some understanding of the information returned in order to properly interpret results.

The real inquiry nature of this software must be designed and modeled by the teacher. Students must have some method of determining which planet to visit by deciding on which ones life might be most likely. If students do not know what the tests are designed to do or what the results mean, the teacher must plan a method for their finding out. As students research and share their knowledge, they perform better on subsequent missions. The class can use the data gathered by all teams working on the project, to determine the answers to some of the problems deductively. Certainly, visiting these planets via the computer promotes many different types of learning and is far more motivating than a worksheet based on the nine planets around the sun.

SIR ISAAC NEWTON'S GAMES (Sunburst) is a set of experiments dealing with the motion of objects (Figure 6.10). The program, designed so that even young children can develop some intuitive ideas about the factors that govern motion, can be used from fifth grade through college. The program contains five parts that introduce students to Newtonian dynamics. Students move a marker in a specific environment (the Earth, near the sun, or out in space), with or without friction acting on its motion, by either kicking the object or letting it coast. The computer immediately shows the results of the action. Students can try this on a track or an obstacle course, in a race with another student, or in playing tag, or in trying to write their names using the skills they have learned in these games.

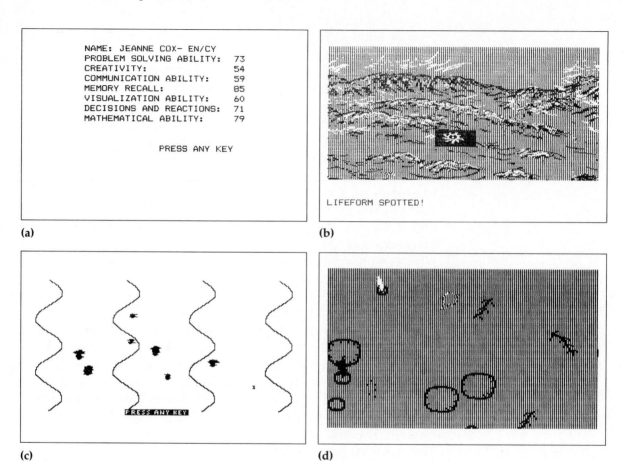

NAME: JEANNE COX- EN/CY
PROBLEM SOLVING ABILITY: 73
CREATIVITY: 54
COMMUNICATION ABILITY: 59
MEMORY RECALL: 85
VISUALIZATION ABILITY: 60
DECISIONS AND REACTIONS: 71
MATHEMATICAL ABILITY: 79

         PRESS ANY KEY

(a)

LIFEFORM SPOTTED!

(b)

PRESS ANY KEY

(c)

(d)

Figure 6.9 In STAR SEARCH, the student selects crew members (screen a). Screen (b) views the planet from its surface. Screen (c) simulates a magnetometer scan from the atmospheric probe. Screen (d) shows an optical telescope scan over the planet from the atmospheric probe.

The programs described in this section are designed to teach scientific problem solving within an inquiry framework in units of instruction related to biology, chemistry, physics, and Earth science. These programs, and others like them, are effective if teachers carefully integrate them with good units of instruction based on the learning cycle principles and incorporate the principles discussed in Chapter 9. Teachers should not simply tell their students to "go over there and run these programs" any

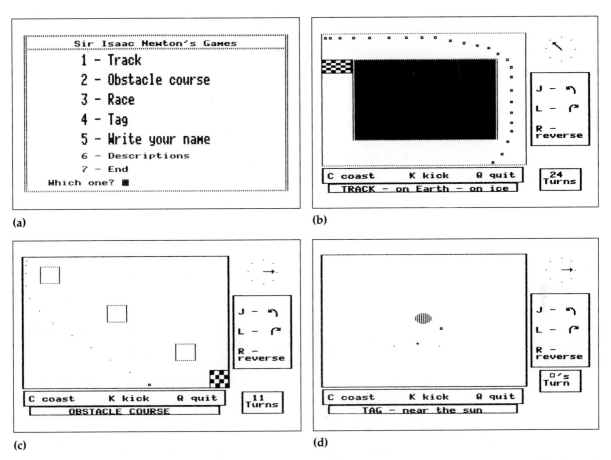

(a)

(b)

(c)

(d)

Figure 6.10 Screen (a) from SIR ISAAC NEWTON'S GAMES shows the menu of the games available. Screen (b) shows the ball being kicked around a track of ice on earth, following Newtonian laws of motion. Screen (c) shows the ball being kicked around obstacles. Screen (d) shows a game of tag (student versus computer) near the sun.

more than they would tell them to use microscopes or any other noncomputerized tool without the appropriate instructional strategies. By providing opportunities that would not be available in the absence of the computer, these programs can greatly enhance academic learning time devoted to learning scientific thinking skills. In addition, teachers should be careful to incorporate the transfer strategies described next.

TRANSFERRING SKILLS

The only reason for developing thinking skills is to use them to think effectively. Students must apply these skills both in academic areas and in settings outside of school. The rest of this chapter examines the relationship of the computer to higher-order thinking skills in specific content areas of the science curriculum.

Regarding the transfer of thinking skills, it is useful to examine two distinct but related topics. First, we look at strategies to help students transfer into a content area thinking skills learned in other settings. These other settings could include other thinking skills programs like those discussed earlier, other materials specifically designed to teach thinking skills, or units of instruction in a different subject matter, such as some area of the science curriculum. Second, we discuss computerized materials that introduce or teach higher-order thinking skills in the context of a specific area of the science curriculum.

Some learners easily see the similarities among various problem-solving settings and almost automatically incorporate into new settings those skills that are relevant from other situations. Students can transfer thinking skills from one area to another just as athletes transfer physical skills from one sport to another. A major difference between physical and intellectual skills, however, is that thinking is an internal process. Learners can directly view experts performing physical activities and determine what skills are needed and when and how to apply them; but it is not always obvious what thinking skills are needed to perform an intellectual task, such as solving a problem. Nor is it easy for a teacher to determine what thinking skills learners already possess in order to plan lessons.

Teachers can help students generalize skills and strategies by employing the "Remember when . . . now let's" strategy. When students are working on a problem in which they could apply a skill or strategy learned in a previous situation, the teacher can simply say, "Remember when we discovered that . . ." and briefly summarize whatever needs to be generalized, helping students to use the visual, auditory, and kinesthetic images from the current and previous learning experiences. The teacher then prompts the learners to try to apply that skill in the new setting: "Now let's look for a way to apply that same strategy here." The following pages present examples of the application of this kind of transfer of learning.

THE FACTORY (see Figure 6.3) can help students develop a wide range of cognitive skills and strategies, depending on how the teacher uses the program. The following examples focus on the strategy of first gener-

Do-It-Yourself Simulations

Most simulations described in this book are "canned" programs based on mathematical models of scientific principles. It is useful for students to run these predesigned programs and to understand the mod- els and principles upon which they are built. Even more useful is the opportunity for students to *build* their own models. The computer can make it easy for students to build models and to focus on useful

Screen (a) shows an imperfect, first try at a model of the deer/ wolf population in a meadowlike environment. It does not take into consideration the fact that a wolf must eat a certain number of deer each year to survive, and it lets the deer population mysteriously drop below zero. Screen (b) improves the model by not permitting the population to drop below zero.

```
039--join(pwolf,pdeer)
        23    1500
        37    1958
        59    2211
        94    1871
       151     738
       241    -243
       386     361
       617   -1206
----------------------PACIFIC CREST SOFTWARE--PC:SOLVE-------------------
026I wolf and deer populations taking into account the predator relationship.
027I Deer deaths from wolves are taken into account.  However, the model
028I does not take into account the maintenance level of wolves.  Also, the
029I deer population is unrealistically allowed to go below zero.
030I years=20
031I pmead=.008
032I deerbr=.6  g wolfbr=.6
033I pwolf=fill(?,years)  g pdeer=fill(?,years)
034I pwolf[1]=23  g pdeer[1]=1500
035I
036I for i=2 to years do deerdeath=(pwolf[i-1] * pdeer[i-1] * pmead)  g deerpop=
037I pdeer[i-1] - deerdeath  g pdeer[i]=deerpop + (deerpop*deerbr)  g pwolf[i]=
038I pwolf[i-1] + (pwolf[i-1] * wolfbr)
039I join(pwolf,pdeer)
```

(a)

```
039--join(pwolf,pdeer)
        23    1500
        37    1958
        59    2211
        94    1871
       151     738
       241       0
       386       0
       617       0
----------------------PACIFIC CREST SOFTWARE--PC:SOLVE-------------------
026I wolf and deer populations taking into account the predator relationship.
027I Deer deaths from wolves are taken into account.  However, the model
028I does not take into account the maintenance level of wolves.  Also, the
029I deer population is unrealistically allowed to go below zero.
030I years=20
031I pmead=.008
032I deerbr=.6  g wolfbr=.6
033I pwolf=fill(?,years)  g pdeer=fill(?,years)
034I pwolf[1]=23  g pdeer[1]=1500
035I
036I for i=2 to years do deerdeath=(pwolf[i-1] * pdeer[i-1] * pmead)  g deerpop=
037I pdeer[i-1] - deerdeath  g pdeer[i]=deerpop + (deerpop*deerbr)  g pwolf[i]=
038I pwolf[i-1] + (pwolf[i-1] * wolfbr)  g pdeer[i]=replace[i],1,0,0)
039I join(pwolf,pdeer)
```

(b)   *(Continued)* problem solving without the needless drudgery that might occur with noncomputerized attempts at model building.

BALANCE (Diversified Educational Enterprises) (described in Chapter 5) teaches students about the predator/prey relationship by enabling them to conduct experiments that manipulate such variables as the number of deer and wolves, the type of environment, and the degree of permissible hunting. Students enter variables, and the program applies a mathematical formula based on a theoretical model to indicate what happened to the deer

Screen (c) adds the factor that wolves must eat an average of twenty-six deer each year to survive. (The formula for "population" referred to in line 36 is automatically generated by the program.) Screen (d) shows the output of the model developed in screen (c).

```
039--join(pwolf,pdeer)
        23      1500
       121      1958
         7       112
        10       170
        15       249
        22       350
        28       463
        35       572
----------------------PACIFIC CREST SOFTWARE--PC:SOLVE-------------------
026| wolf and deer populations taking into account the predator relationship.
027| Deer deaths from wolves are taken into account.  The model now
028| does take into account the maintenance level of wolves.  Also, the
029| deer population is unrealistically allowed to go below zero.
030| years=20
031| pmead=.008
032| deerbr=.6   & wolfbr=.6
033| pwolf=fill(?,years)   & pdeer=fill(?,years)
034| pwolf[1]=23   & pdeer[1]=1500
035| maintlev=26
036| for i=2 to years do population
037|
038|
039| join(pwolf,pdeer)
```

(c)

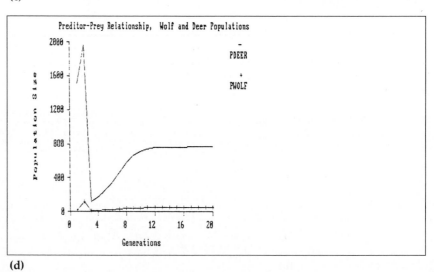

(d)

and wolf populations as a result of the manipulation.

BALANCE is an excellent program, but students can develop and test the same model as BALANCE by using a problem-solving program. With a problem-solving program, students supply the model, and the computer shows the results. If the results are not satisfactory, students can modify the model until they obtain appropriate outcomes. A program like PC:SOLVE (Pacific Crest Software) supplies the tools that students and teachers need to build and test models in a wide variety of domains, including science.

Besides helping students understand the principles behind the predator/prey relationship, a program like PC:SOLVE lets them focus on important "what if" questions. What if an additional factor, such as variations in birth rate of the animals or fluctuations in temperature, were added to the model? With BALANCE, students can (and should) talk about the impact of these "what if" questions, but PC:SOLVE lets them add these factors to the model and test the revision. Because BALANCE is written in BASIC, modification of that program is also possible—but only by searching out the proper lines in the program, altering the program code, and hoping that the changes have no adverse influence on the operation of the rest of the program. With copy-protected programs, students or teachers cannot alter the model. With a program like PC:SOLVE, modification, expansion, and enhancement of models are simple and natural accomplishments.

This kind of program can be used either in large-group, small-group, or individual presentations, as described in Chapter 9. PC:SOLVE provides myriad functions and procedures that perform mathematical and statistical calculations, graph functions, and manage data to help students solve problems, develop models, and write reports. Many of these procedures are more complex than would be required for precollege students: but in that case, teachers can simply ignore the more sophisticated procedures and let their students use the computer to perform tasks appropriate to their educational needs.

A tool program like PC:SOLVE lets students explore concepts that would be difficult to pursue if the calculations had to be done by hand or through separate computer programs. The program puts into a single package nearly all the tools that a science student could need and makes these tools easily accessible for problem solving. The program can also be used in many areas across the curriculum. Its publisher offers materials that suggest using this program as part of a separate unit on problem solving and critical thinking and then generalizing these higher-order thinking skills to various curriculum areas, including science, mathematics, and social studies.

Both BALANCE and PC:SOLVE can play a viable role in the science curriculum. The teacher needs to examine both kinds of programs and decide which is appropriate for a specific educational context.

ating several solutions, then evaluating them to determine which one is best. In these examples, the teacher makes the statements in the following dialogue and then branches to the appropriate situation.

TEACHER: *Remember when* we found many different solutions to the problem in THE FACTORY? What did we do?

STUDENTS: We generated as many solutions as we could. Then we set up criteria, evaluated each of our solutions, and chose the one that was the best.

TEACHER: *Now let's* apply the same strategy to this problem.

1. Biology: "Let's get into our small groups and think of as many different ways as possible to measure blood volume. Start by having each person in the group think of several ways. Then discuss each way with the group to refine it and make it as clear as possible. Then we'll compare the ideas from all the groups, and finally, I may show you some other ways we could use. Afterward, we'll vote to see which one is best. Remember, there will be many good ways; but try to develop some that are clear, accurate, and easy to use."

2. Discipline problem: "We really cannot have disturbances like this in this class. I understand why you're acting this way; but there must be a solution. In fact, there must be at least ten different ways to solve this problem. We're going to discuss this problem after school today until we find those ten ways, and then we're going to choose and implement the strategy that we think will best solve the problem."

3. Industrial education: "I want you to use the CAR BUILDER program to develop an automobile that will get at least thirty miles per gallon, will seat four people, and will have a cruising range of at least four hundred miles. In addition, I want you to incorporate features that make it attractive to a buyer. Remember what we did in THE FACTORY. First, design a few cars that meet the standards and then see what you can do about developing a best strategy to make your car both efficient and attractive."

4. Geometry: "Here is a geometric figure. I want you to determine the area of triangle *XPR*. There are several different ways to do this. First, I want you to find as many distinctly different ways as possible to determine the area. Then I want you to select the one that will accomplish the task with the fewest computational steps."

The preceding learning situations are extremely diverse. Only Item 1 is directly related to science. Item 2 would probably occur in a science classroom, but most students would not automatically relate solving personal problems to scientific thinking. Items 3 and 4 would occur in classrooms somewhere else in the school; and to promote transfer, the science teacher might have to take steps to share ideas with the teachers of these other subjects. What all these situations have in common is that they all apply one of the instructional strategies that students can employ in running THE FACTORY. Note that only one of them (Item 3) directly uses the computer. In another (Item 4), students could use a computer program like GEOMETRIC SUPPOSER (Sunburst) to solve the problem, but this would not be necessary. In the other two, it is improbable that students would use computers during this generalization lesson. In each case, however, the teacher is helping students apply to a new problem strategies that they learned or practiced while running THE FACTORY.

In some cases, the application of skills from a program like THE FACTORY to a specific subject area may be even more direct. For example, THE FACTORY also teaches students to plan ahead and to use concrete models to simplify abstract processes. Real factories sometimes fail because their owners and operators neglect to apply these strategies. If students are reading about such an instance in science or social studies, it may be useful to let them run or rerun THE FACTORY during that unit of instruction to let them see the connection between their own experience and the historical events.

Although the preceding examples are stated as if the teacher who is presenting the generalization lesson is the same teacher who used THE FACTORY with the students, this is not a necessity. What is important is that the teacher who is trying to make the generalization must be aware of the thinking strategies employed in THE FACTORY and able to show students how these relate to the current learning situation. In other words, the first paragraph can be modified to say, "Remember last year when you found many different solutions to the problem in THE FACTORY? What did you do?" or "Remember when you ran THE FACTORY in your science class with Mr. Jones . . . ?"

Also note that the exact structure of the generalization lesson may vary. Students who have become very adept at applying the strategy may need no more than the introductions stated in the preceding examples. In other cases, the teacher may wish to give continued guidance. For example, in the geometry example (Item 4), the teacher might put GEOMETRIC SUPPOSER on a large screen and lead a group discussion to solve the problem, serving as a model and mediator to help students reapply strategies that they had previously learned. In addition, the teacher could use a very nonexpository approach. For example, the teacher might say, "Does this remind you of any program we've run recently?" or "How is what we're doing now similar to what we did in THE FACTORY?" The critical point is that students will transfer a skill or strategy more readily if they have it specifically in mind at the time they need it in the new situation.

Finally, it is important to note that the transfer of this skill or strategy need not be the sole focus of what we have called the generalization lesson. In the preceding examples, students and teachers would be mainly interested in the contents covered by the topic of discussion. The strategy being generalized from THE FACTORY would be a secondary objective. To help students meet multiple objectives, simply focus their attention on the various objectives under consideration. For example, in the biology lesson (Item 1), the teacher would probably begin the lesson by saying, "Standardized methods of measurement enable us to compare two or more items. A good method of measurement should be clear, accurate, and easy to use." Students would be pursuing four objectives during this unit of instruction:

1. Defining and giving examples of measurement
2. Identifying the characteristics of good methods of measurement
3. Making accurate measurements using the instruments designated by the teacher or selected by students
4. Generating several solutions to an "open ended" problem and then selecting the best solution by comparing each to a set of specified criteria.

It is possible that a unit of instruction could become cluttered with an excessive number of objectives, and the result would be confusion and faulty learning among students. However, this is not likely to occur in the example we are discussing because the objectives are interrelated. The first two are lower-order content-area objectives. The third is a higher-order content-area objective. The fourth is a noncontent-area objective, but it serves as the means by which students meet the other three objectives. By simply focusing the students' attention on this fourth objective, the teacher helps them generalize this important skill without detracting from the other objectives, which would normally be covered in this lesson.

Besides helping students transfer skills learned in general thinking skills programs into the science curriculum, it is possible to use the computer to teach higher-order thinking skills in the context of specific science topics. In such cases, the unit of instruction has a dual purpose: (1) to teach something about the subject matter and (2) to teach the process skills under consideration. By appropriately focusing attention, the teacher can enable students to master both sets of objectives.

For example, BALANCE (described in greater detail in Chapter 5) enables students at various grade levels to study the predator/prey relationship in nature. Students who run this program can come away with such factual, biological understandings as: "There must be a balance between predator and prey. When too many predators die, this is harmful for the prey. And if there are too many predators, they will eat all the prey, and then the predators will starve too." These are important insights, and they can help students become more intelligent citizens. However, while running this program, students conduct experiments; and if they focus their attention appropriately, they can develop or apply scientific *process* skills as well. To help teach these process skills, the teacher should direct students' attention to such issues as (1) it is best to manipulate just one variable at a time and (2) data must be tabulated in an effective manner if they are to be easily interpreted.

COMBINING THE COMPUTER WITH OTHER APPROACHES

Of course, it is not really necessary to use *either* a content program *or* a thinking skills program—it is possible to use bo*th*. The teacher could use THE INCREDIBLE LABORATORY first, then BALANCE, and then a video segment from "Thinkabout." Properly used, all three sets of materials require the generation and testing of hypotheses in a scientific manner to solve a problem. Furthermore, each set of materials permits multiple solutions to a problem, and students must evaluate alternate solutions according to stated criteria to determine which is best. A good teacher would use these and other materials in conjunction with the "remember when" rule.

Our belief is that it is seldom, if ever, a good idea to use a series of computer programs to replace the classroom teacher. Rather, the computer program is used as a tool in an instructional sequence coordinated by the teacher or, in the case of independent learning, coordinated by the learner. Two major roles of the teacher are to serve as a model and as a mediator. As a model, the teacher should use effective thinking skills that are appropriate to the learning level of students. The teacher should often verbalize these skills or in some other way direct students' attention to the usefulness of these skills. The teacher's role as a mediator includes helping (forcing) students to direct their attention to those areas of the program or problem where it should effectively be directed.

SUMMARY

This chapter has focused specifically on the thinking skills aspect of science programs. In science classes, students can learn thinking skills that they can generalize not only to other science classes but also to many other curriculum areas and real-life situations. These skills include abilities to observe and record information, classify, communicate results, draw inferences, and generalize conclusions, as well as an appreciation of the scientific method and the overall activity of scientists. These skills are usually taught most effectively within an inquiry framework based on the principles of the learning cycle. This chapter has shown how computer programs, usually integrated with teacher guidance and other strategies, can help students develop and generalize these skills.

REFERENCES

Educational Technology Center. *Making Sense of the Future*. Cambridge, Mass.: Harvard Graduate School of Education, 1988.

Friedler, Y., R. Nachmias, and M. C. Linn. *Using Microcomputer-Based Laboratories to Foster Scientific Reasoning Skills*. Berkeley: University of California, Computer as Lab Partner Project, 1987.

Mallan, J. T., and R. Hersch. *No G.O.D.'s in the Classroom: Inquiry into Inquiry*. Philadelphia: Saunders, 1972.

Rogers, C. *On Becoming a Person*. Boston: Houghton Mifflin, 1961.

Rosenshine, B., and N. Furst. "Research on Teacher Performance Criteria." In B. O. Smith (Ed.), *Research in Teacher Education*. Englewood Cliffs, N.J.: Prentice-Hall, 1971.

Vockell, E., and R. van Deusen. *The Computer and Higher-Order Thinking Skills*. Watsonville, Calif.: Mitchell, 1989.

CHAPTER 7

USING THE COMPUTER AS A TOOL IN SCIENCE

ONE RESULT OF THE influx of microcomputers into the educational setting has been the increased productivity experienced by both teachers and students. The computer can be used for a wide variety of general-purpose tasks which help users be more productive in any area.

WORD PROCESSING PROGRAMS

Many word processing programs are available for classroom use. Vockell and Schwartz (1988) discuss these programs and related software (including spelling checkers, thesauri, outlining programs, and text analysers). Schwartz and Vockell (1989) discuss specific uses of these programs to help students develop writing skills.

Both teachers and students in science classes should use the functions of a word processing program. Word processing allows a person to type, edit, save, and print text information. Teachers can write letters, type lesson plans and tests, and prepare a variety of material for class use. Students can write their lab reports and prepare their science fair projects on the computer. By using an integrated program like APPLEWORKS, users combine word processing capabilities with database management strategies to customize labels, create feedback letters containing students' grades, generate term papers containing the results of a database search, and send orders to supply companies.

The term *template* refers to forms that are saved without additional information put in them so they can be used repeatedly. For example, a teacher might set up the margins, indenting structure, and question format for a test and save it as a template. When it is time to prepare a test, the template can be called up, the name changed, and the test prepared on the form. The form is still present on the disk under the old name. A template for a student laboratory report can be created in the same manner. This enables all students to submit reports in a standardized format without having to engage in the nonscience-related, comparatively trivial task of aligning margins and the like. Templates make it easier for students to include all information that is supposed to go into a report. Many public domain and commercial disks with various templates for teachers and for science curricula are available. For example, K–12 Micromedia publishes the SCHOOLWORKS series, which includes such templates as "Athletic Director," "Letter Files," "Media Center," "Office," "Social Studies," and "IEP" (individualized educational program). Figure 7.1 shows a laboratory template for a "Shadow Lab" from a public domain disk. In this case, the

Figure 7.1 "Shadow Lab," an APPLEWORKS word processing file, is printed with TIMEOUT SUPER-FONTS (Beagle Brothers).

Name................................   Date............
Length of a Shadow Lab   Period...........

Question: How does the length of a shadow vary during the day?

Introduction: As you know, the sun's position changes during the day as well as during the year. This activity will better help you to understand the daily (diurnal) movement of the sun, and its relationship to the length of a shadow.

Materials: <u>Meter Stick</u>, Sun, Graph Paper, & pencil

Method: 1. Record the length of a shadow cast by a meter stick from sunrise to sunset on a single day. You should make at least 3 morning, 3 afternoon readings, and a reading near noon. Record these lengths in meters on a data table.

2. Graph your results on a piece of graph paper, using the horizontal axis to graph the constantly changing variable (time), and the vertical axis to represent the length of the shadow in meters.

3. Answer the following questions using complete sentences.

Questions:
1. Describe the changing length of the shadow from sunrise to sunset.

2. What is the relationship between the altitude of the sun and the shadow's length? Is this a direct or inverse relationship?

3. Why do we call the sun's path an apparent path?

4. Where must the sun be if the length of the shadow was 0 meters?

5. Compare the length of the sun's shadow at noon next month, in December, and in June with today's measurement.

6. For an observer in NYS, in what compass direction would you find the sun at solar noon?................ In what direction would the shadow be cast by the noon sun?.............

template actually serves as a worksheet. Students load the template into their computers and use the word processor to type the answers after the questions. Needless typing is eliminated. Teachers often acquire a template like this and then modify it slightly to suit the needs of their own class.

Public domain template disks are available from several sources, including the following:

- TI&IE (Teachers' Idea & Information Exchange), Dept. TW, PO Box 6229, Lincoln, NE 68506
- National AppleWorks Users Group (NAUG), Box 87453, Canton, MI 48187
- Education Disk Resources for AppleWorks, PO Box 24146, Denver, CO 80224
- Resources for AppleWorks collection from The Works. The Works, PO Box 72, Dept. 37, Leetsdale, PA 15056–0072.
- The AppleWorks Users Group (TAWUG), PO Box 24789, Denver, CO 80224

APPLEWORKS lets the user move from one application to the other. For example, the teacher could create a customized letter for parents as shown in Figure 7.2. If the database contains the names and addresses of parents and students, grades, and personal information about each student, the MAIL MERGE function allows the teacher to merge this information into a personalized letter.

SPREADSHEET PROGRAMS

The electronic spreadsheet allows the user to calculate and recalculate numerical data. This capability lets a teacher keep easy track of all student grades. The computer can calculate all types of statistics—totals, averages, mean, standard deviation, and others. Spreadsheets can hold inventories of chemical and other science supplies and the cost of these items in order to manage the science department budget. The recalculation capability of the spreadsheet makes it easy to test a hypothesis by examining what would happen if the value of one or another variable in an experiment were changed. Other uses of the spreadsheet allow users to examine mathematical formulas to see how changing one variable may influence others in the equation. Finally, many experiments requiring the manipulation of data can be entered on the spreadsheet, with the computer filling in the calculated columns.

Templates for grades and science experiments are available. One valuable source is MECC's APPLEWORKS SPREADSHEETS FOR MATHEMATICS AND SCIENCE. Some spreadsheets include look-up tables, which let teachers assign grades to certain scores obtained by students, without constant reference to the manual. Another application is in the Science Olympiad event "Astronomy," in which students participate

Figure 7.2 Screen (a) shows a template in which all database fields are merged with the form letter.

```
                                                    October 5, 1990



    Mr. and Mrs. ^<PNAME> ^<PLNAME>
    ^<ADDRESS>
    ^<CITY>, ^<STATE>  ^<ZIP>



    Dear Mr. and Mrs. ^<PLNAME>,


        I have included your ^<RELATION>'s grades for the first
    month of school.  The assignment is given first, then the
    number of points possible (in parentheses), and that is
    followed by your ^<RELATION>'s numerical grade on that
    assignment.

    Letter to Congressman (10):     ^<LETTER (10)>
    Questions on page 10 (5):       ^<QUESTIONS1 (5)>
    Quiz over Chapter 1 (10):       ^<QUIZ (10)>
    Lab 1 (10)                      ^<LAB 1 (10)>
    Lab 2 (10)                      ^<LAB 2 (10)>
    Quiz over Chapter 2 (20)        ^<QUIZ (20)>
    Questions on page 41 (5)        ^<QUESTIONS2 (5)>
    Unit Test (100)                 ^<TEST (100)>


    The total points possible are:  ^<POSSIBLE>
    Your son's total so far is:     ^<TOTAL>
    The grade at this time would be: ^<GRADE> %


        Should you have a question about missing or poor work,
    please call me at the school.



                                        Sincerely,



                                        Mrs. Betsy Ross

                                        7th grade
                                        science teacher
```

(a)

in a practical identification test in a planetarium and in an outdoor skill test to locate a designated object with a homemade astrolabe. Instead of calculating points for performance by hand, scorekeepers enter performance data into a spreadsheet program. Figure 7.3 shows part of this spreadsheet for a recent state Science Olympiad.

Any science experiment that involves calculations can be done on a spreadsheet set up before the experiment begins. Figure 7.4 shows a spreadsheet used to calculate the relationship between various independent variables and the period of a pendulum from data collected by a

Figure 7.2 continued
Screen (b) shows an example
of a form letter to parents.

```
                                                    October 5, 1990



        Mr. and Mrs. John J. Adams
        101 Third Avenue
        Earlytown, KS  66600



        Dear Mr. and Mrs. Adams,

            I have included your son's grades for the first month
        of school.  The assignment is given first, then the number
        of points possible (in parentheses), and that is followed by
        your son's numerical grade on that assignment.

        Letter to Congressman (10):       8
        Questions on page 10 (5):         4
        Quiz over Chapter 1 (10):         6
        Lab 1 (10)                        6
        Lab 2 (10)                        7
        Quiz over Chapter 2 (20)
        Questions on page 41 (5)
        Unit Test (100)                  64

        The total points possible are:   170
        Your son's total so far is:       100
        The grade at this time would be: 58.82 %

            Should you have a question about missing or poor work,
        please call me at the school.




                                                    Sincerely,



                                                    Mrs. Betsy Ross

                                                    7th grade
                                                    science teacher
```

(b)

student in a laboratory. The only numbers that the student adds to the spreadsheet are the periods of the pendulum at various lengths.

Pogge and Lunetta (1987) describe several spreadsheet examples based on material from the BSCS green version text, *Biological Science: An Ecological Approach,* which show population sampling, population simulation, and heights of plants at the top, middle, and bottom of slopes on which a two-way analysis of variance was conducted. Note, however, that students

(a)

```
File:    STATE.SO
                  Altitude=  7.50  Azimuth=  289.00                    12.00
Num.        School Particip  Alt.      Alt      Az       Az     Alt     Az    Constr    Total  TOTTotal
                                       Diff             Diff   Points  Points  Points   AstroIPts deduct
```

Num.	School	Particip	Alt.	Alt Diff	Az	Az Diff	Alt Points	Az Points	Constr Points	Total AstroIPts	TOTTotal deduct
1	BLAIR MS		8.00	.50	292.00	3.00	14.00	13.00	7.00	34.00	
2	WEST MS		6.00	1.50	272.00	17.00	14.00	8.00	9.00	31.00	
3	McAULIFFE JH		18.00	10.50	285.00	4.00	5.00	12.00	3.00	20.00	13.00 used compass
4	RESNIK MS		8.00	.50	288.00	1.00	14.00	14.00	8.00	36.00	
5	OAKDALE MS		9.00	1.50	240.00	49.00	14.00	5.00	5.00	24.00	
6	BLUE RIVER JH		8.00	.50	283.00	6.00	14.00	11.00	8.00	33.00	
7	MUIR MS		13.00	5.50	270.00	19.00	10.00	8.00	7.00	25.00	
8	POWELL MS		9.00	1.50	300.00	11.00	14.00	9.00	9.00	32.00	
9	MOUNTAIN VIEW MS		10.00	2.50	295.00	6.00	13.00	11.00	10.00	34.00	
10	SCOBEE JH		6.00	1.50	196.00	93.00	14.00	3.00	6.00	23.00	
11	VALLEY VIEW JH		6.80	.70	293.00	4.00	14.00	12.00	11.00	37.00	
12	DEEP SPRINGS MS		5.00	2.50	308.00	19.00	13.00	8.00	7.00	28.00	
13	ONIZUKA MS		6.50	1.00	275.00	14.00	14.00	9.00	5.00	28.00	
14	JARVIS JH		7.50	0.00	278.00	11.00	14.00	9.00	5.00	28.00	
15	EISENHOWER JH		9.00	1.50	2.00	287.00	14.00	1.00	4.00	19.00	12.00 used compass
16	KENNEDY MS		9.00	1.50	293.00	4.00	14.00	12.00	11.00	37.00	
17	KING MS		6.00	1.50	288.00	1.00	14.00	14.00	8.00	36.00	
18	CARVER MS		6.50	1.00	288.00	1.00	14.00	14.00	12.00	40.00	
19	LINCOLN JH		7.00	.50	285.00	4.00	14.00	12.00	8.00	34.00	
20	WASHINGTON JH		6.00	1.50	266.00	23.00	14.00	7.00	5.00	26.00	
21	POWHATTAN MS		98.00	90.50	61.00	228.00	1.00	1.00	10.00	12.00	
22	SMITH JH		8.50	1.00	289.00	0.00	14.00	14.00	9.00	37.00	
23	COLUMBUS JH		5.50	2.00	285.00	4.00	13.00	12.00	7.00	32.00	
24	LEIF ERICKSON MS		12.00	4.50	286.00	3.00	11.00	13.00	4.00	28.00	21.00 used compass
25	McNAIR JH		13.00	5.50	272.00	17.00	10.00	8.00	6.00	24.00	
26	WHITE PINE MS		8.00	.50	71.00	218.00	14.00	1.00	7.00	22.00	
27	BYRD JH		7.00	.50	286.50	2.50	14.00	13.00	11.00	38.00	
28	SAN MARTIN MS		11.00	3.50	106.00	183.00	12.00	1.00	6.00	19.00	
29	MERIDIAN JH		6.00	1.50	290.00	1.00	14.00	14.00	5.00	33.00	
30	PARALLEL MS		7.00	.50	291.00	2.00	14.00	13.00	10.00	37.00	
31	SUPERIOR MS		0	7.50	0.00	289.00	0.00	0.00	0.00	0.00	
32	PACIFIC JS		7.00	.50	60.00	229.00	14.00	1.00	8.00	23.00	
33	ATLANTIC MS		83.10	75.60	87.50	201.50	1.00	1.00	6.00	8.00	
34	GREEN HILLS MS		5.50	2.00	299.50	10.50	13.00	9.00	11.00	33.00	
35	GRAND BAY JH		7.00	.50	265.00	24.00	14.00	7.00	8.00	29.00	
36	IROQUOIS MS		1.00	6.50	1.00	288.00	9.00	1.00	0.00	1.00	participation
37	METROPOLIS JH		11.00	3.50	291.00	2.00	12.00	13.00	10.00	35.00	
38	BALBOA JH		8.00	.50	303.00	14.00	14.00	9.00	9.00	32.00	
39	CORONADO MS		1.00	6.50	278.00	11.00	9.00	9.00	7.00	25.00	
40	FRANKLIN MS		7.00	.50	39.00	250.00	14.00	1.00	5.00	20.00	

```
            lookup  tableAltitu
 0.00        1.00   2.00  3.00   4.00    5.00    6.00    7.00    8.00    9.00   10.00   20.00  30.00  40.00  50.00  500.00
14.00       14.00  13.00 12.00  11.00   10.00    9.00    8.00    7.00    6.00    5.00    4.00   3.00   2.00   1.00    1.00


            lookup  tableAzimut
 0.00        1.00   2.00  4.00   6.00    8.00   10.00   15.00   20.00   25.00   30.00   50.00  90.00 120.00 180.00 360.00
14.00       14.00  13.00 12.00  11.00   10.00    9.00    8.00    7.00    6.00    5.00    4.00   3.00   2.00   1.00    1.00
```

Figure 7.3 Screen (a) shows a spreadsheet that accepts raw data supplied by the contestants, calculates the difference in altitude and azimuth, and assigns point values based on performance. Figure (b) is the printout using TIMEOUT SIDESPREAD (Beagle Brothers).

(b)

```
File: STATE.SO                   REVIEW/ADD/CHANGE              Escape: Main Menu
=========E=========F=========================G=====================H=============
  1|  Azimuth=   289
  2|    Alt        Az             Az                          Alt
  3|    Diff        0            Diff                        Points
  4|
  5|@ABS(D5-D1)    292   @ABS(F5-F1)    @IF(D5>0,@LOOKUP(E5,A48...P48),0)
  6|@ABS(D6-D1)    272   @ABS(F6-F1)    @IF(D6>0,@LOOKUP(E6,A48...P48),0)
  7|@ABS(D7-D1)    285   @ABS(F7-F1)    @IF(D7>0,@LOOKUP(E7,A48...P48),0)
  8|@ABS(D8-D1)    288   @ABS(F8-F1)    @IF(D8>0,@LOOKUP(E8,A48...P48),0)
  9|@ABS(D9-D1)    240   @ABS(F9-F1)    @IF(D9>0,@LOOKUP(E9,A48...P48),0)
 10|@ABS(D10-D1)   283   @ABS(F10-F1)   @IF(D10>0,@LOOKUP(E10,A48...P48),0)
 11|@ABS(D11-D1)   270   @ABS(F11-F1)   @IF(D11>0,@LOOKUP(E11,A48...P48),0)
 12|@ABS(D12-D1)   300   @ABS(F12-F1)   @IF(D12>0,@LOOKUP(E12,A48...P48),0)
 13|@ABS(D13-D1)   295   @ABS(F13-F1)   @IF(D13>0,@LOOKUP(E13,A48...P48),0)
 14|@ABS(D14-D1)   196   @ABS(F14-F1)   @IF(D14>0,@LOOKUP(E14,A48...P48),0)
 15|@ABS(D15-D1)   293   @ABS(F15-F1)   @IF(D15>0,@LOOKUP(E15,A48...P48),0)
 16|@ABS(D16-D1)   308   @ABS(F16-F1)   @IF(D16>0,@LOOKUP(E16,A48...P48),0)
 17|@ABS(D17-D1)   275   @ABS(F17-F1)   @IF(D17>0,@LOOKUP(E17,A48...P48),0)
 18|@ABS(D18-D1)   278   @ABS(F18-F1)   @IF(D18>0,@LOOKUP(E18,A48...P48),0)
--------------------------------------------------------------------------------
H1

Type entry or use @ commands                                     @-? for Help
```

```
Type in period for each pendulum on the table below.

To calculate  table, press Apple-K (Calculate).

The mathematical relationship between period and pendulum length
can be found by identifying which column of values remains
constant for each of the experimental values of the pendulum
length.
```

		Direct Relationships			Inverse Relationships		
Length	Period	$T = KL$	$T = KL^2$	$T^2 = KL$	$T = K/L$	$T = K/L^2$	$T^2 = K/L$
(L)	(T)	T/L= K	T/(L^2)= K	(T^2)/L= K	TL= K	T(L^2)= K	(T^2)L= K
.2		0.000	0.000	0.000	0.000	0.000	0.000
.4		0.000	0.000	0.000	0.000	0.000	0.000
.6		0.000	0.000	0.000	0.000	0.000	0.000
.8		0.000	0.000	0.000	0.000	0.000	0.000
1		0.000	0.000	0.000	0.000	0.000	0.000
1.2		0.000	0.000	0.000	0.000	0.000	0.000
1.4		0.000	0.000	0.000	0.000	0.000	0.000
1.6		0.000	0.000	0.000	0.000	0.000	0.000
1.8		0.000	0.000	0.000	0.000	0.000	0.000
2		0.000	0.000	0.000	0.000	0.000	0.000
2.2		0.000	0.000	0.000	0.000	0.000	0.000
2.4		0.000	0.000	0.000	0.000	0.000	0.000
2.6		0.000	0.000	0.000	0.000	0.000	0.000
	Avg.	ERROR	ERROR	ERROR	ERROR	ERROR	ERROR
	Min.	0	0	0	ERROR	ERROR	ERROR
	Max.	0	0	0	ERROR	ERROR	ERROR
Average-drop max & min.		0	0	0	ERROR	ERROR	ERROR

```
File:   Pendulum
  The Pendulum Problem
```

Figure 7.4 After the student tests pendulums of several different lengths, enters the period of the pendulum at those lengths into the APPLEWORKS spreadsheet template, and presses open-apple K key, the spreadsheet performs the calculations. This has been printed with TIMEOUT SIDESPREAD.

would learn the most by *defining* or at least understanding the underlying logic of the spreadsheet—rather than just plugging information into cells without understanding why they are doing so. The electronic spreadsheet should serve as a tool to minimize unproductive work, not as a crutch. Students still must understand the scientific principles underlying what they are doing.

DATABASE MANAGEMENT PROGRAMS

A database program keeps inventory lists, names and addresses of science suppliers, results of trials in an experiment, or student grades on quizzes, exams, projects, and homework. Database programs also make it easy to compile and retrieve data on a topic; for example, a file might contain extensive information about dinosaurs or student notes about common living organisms.

The essential characteristics of a good database are the quality of the information contained in it and the ease and accuracy with which this information can be retrieved. Because careful observation, coding, and record keeping are important skills of successful science students, the computerized database often plays a useful role in the science classroom. Students normally learn first to use existing databases, then to add information to existing databases, and finally to develop and use databases of their own. Vockell and Schwartz (1988) describe in greater detail database management programs and effective strategies for using them.

One of the critical skills in hypothesis testing is the ability to compile, analyze, and draw conclusions from a set of data. All too often, this data set is very small because of time and resource constraints or because of the inability of the human to make sense from huge quantities of data. The computer's ability to quickly analyze large amounts of data makes it an ideal tool in many science laboratory experiments.

The microcomputer database offers a useful and instructive tool for the science classroom, allowing students to enter, sort, arrange, select, and organize data based on their own hypotheses. Students can test "what if" questions or use the database to quickly eliminate many unwanted possibilities. The well-designed database eliminates the need for hand-written charts and keys through which a student must carefully trudge to eliminate all wrong turns.

Because specific database applications to the science curriculum are described in detail in Chapters 3–5, additional examples are not provided here.

DATA ANALYSIS PROGRAMS

The real power of a microcomputer lies in its ability to deal with large amounts of data quickly. This computational capacity helps solve major difficulties in teaching mathematical models in various areas of science and in the analysis of large amounts of data. What has often been impractical has become easy with the aid of the microcomputer.

Several software packages allow students to enter data and then have the computer either show appropriate transformations or draw graphs of the data. In addition, graphs can easily be redrawn to more appropriate scales in order to focus more precisely on phenomena under consideration. In many of the programs, data are entered in x,y-pairs, and the computer analyzes the information. To make use of such programs, students still have to understand the basic concepts, such as what the x- and y-variables are and what it means to perform a least squares approximation. In addition, as more students use computers, it will become increasingly important for them to develop estimation skills. Students will benefit from using the computer as a tool in data analysis only if they understand the basic principles involved in the analysis and have a general idea what outcome the computer should produce. Otherwise, the computer simply makes it possible to misunderstand and misapply information more rapidly than would be possible without it. With regard to data analysis, science education overlaps with mathematics instruction. Dinkheller, Gaffney, and Vockell (1989) provide a more complete discussion of the problems inherent in the application of the computer as a tool in mathematics instruction.

DATA ANALYSIS (EduTech) is a commercial program that allows the input of data points in x,y-pairs, which can be stored, graphed, or printed. Data can be saved to disk for future reference. In addition, twelve activities that provide hands-on experiments for students to gather their own data and perform analyses using various transformations are included. For example, students could measure the circumference and the diameter of many round objects and enter the data into a file. The file can then be shown in table format (Figure 7.5) and with various graphs to help students arrive at the conclusion that the ratio between these two variables is a constant.

REVIEW AND TEST PRODUCTION PROGRAMS

As we pointed out in Chapter 2, excellent commercially produced drills and tutorials are not plentiful for science instruction. Because of this scarcity, many science teachers use a computer-assisted instruction (CAI) authoring program to produce their own tutorial on specific concepts. Vockell and Schwartz (1988) describe several of these. DATATECH MENTOR MASTER (DataTech Software Systems) is an authoring program particularly

Figure 7.5 After students measure the diameter and the circumference of various spherical and cylindrical objects such as cans, bottles, and balls, they enter the data points and create the graph of the data.

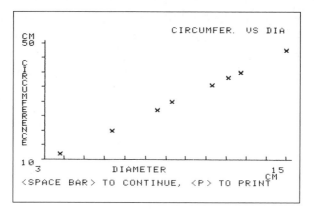

suited to the science curriculum. It is menu-driven, and a teacher can construct information sections, true/false questions, multiple-choice questions, and short-answer questions around commentary and pictures. The questions can be presented in a quiz or a game format. A major strength for the science teacher is that this package includes eight different picture disks with diagrams that can be integrated into the computerized unit. Disks are available on space science, meteorology, plant anatomy, cells/genetics, earthquakes, oceanography, geology, and human anatomy. After the teacher designs an instructional unit, students can receive individual disks tailored to their specific needs. A management component helps the teacher keep track of student performance.

SCIENCE BASEBALL GAMES: EARTH SCIENCE (Create-A-Test) has several categories of questions, with about thirty questions in each category in a baseball game format. If the question is answered correctly, the student gets a hit. The teacher can delete or change questions and add entire categories on the data disk. There are game disks for physics, biology, chemistry, physical science, and elementary science.

Many of the current textbooks used in the science curriculum come with their own test production software and databases. Some of these are good, but many include only very low-level, cognitive questions. The program may only allow multiple-choice, true/false, or fill-in-the-blank questions. Some allow the teacher to input different questions into the question bank, and some allow for machine grading.

A number of companies offer test production programs. For instance, the CREATE-A-TEST (Create-A-Test) system prepares tests for class distribution, supports all question types, and allows teachers to organize their own question banks. This package offers prepared question banks for all science topics for grades 6–12, as well as for several nonscience topics. The built-in editor lets the teacher edit questions and add new ones. (Vockell and Schwartz, 1988, and Vockell and Hall, 1989, discuss test-generating programs in greater detail.) Although programs of this type are useful for very specific, objective questions, they lack flexibility and often do not test higher-level objectives.

OTHER TOOLS

Other programs are available to help with various tasks. PRINT SHOP (Figure 7.6a) and the graphics disks, which accompany it, help the user make graphically illustrated signs, letterheads, or banners. CERTIFICATE MAKER helps design and print certificates of all kinds and for any class or accomplishment. Many teachers use CROSSWORD MAGIC (Figure 7.6b) to prepare crossword puzzles to test vocabulary.

TIMELINER (Tom Snyder Productions) is useful both for science and social studies classes—or for integrating the two fields (Figure 7.7). Students can use timelines that are already on one of five different data disks, including one on science and technology. Students can also make their own timelines, ranging from a very short period of time to the entire history of the Earth, to gain a visual perspective of the relationship between events and a "picture" of the relative amount of time between these events. Timelines are proportionally correct and can be resized to some extent. In addition, the program will print up to ninety-nine pages of timelines.

SCIENCE HELPER K–8 (PC-SIG) is a valuable tool for elementary and middle school science teachers and curriculum writers. This CD-ROM disk offers a menu-driven database of about 1,000 activity-based science and mathematics lessons for grades K–8. All lessons on the disk emphasize the interaction with concrete materials in a laboratory setting. To permit effective retrieval, the lessons are categorized by curriculum project, academic subject, content theme, processes of science, and teacher guide title. In addition, the lessons may be accessed by key words and phrases. For example, a teacher may look for all third-grade lessons in life science that emphasize classification skills. Teachers can obtain printouts of desired lessons with appropriate graphics and modify them as needed for their own units of instruction.

Figure 7.6 A safety sign for the science classroom prepared with PRINT SHOP (screen a).

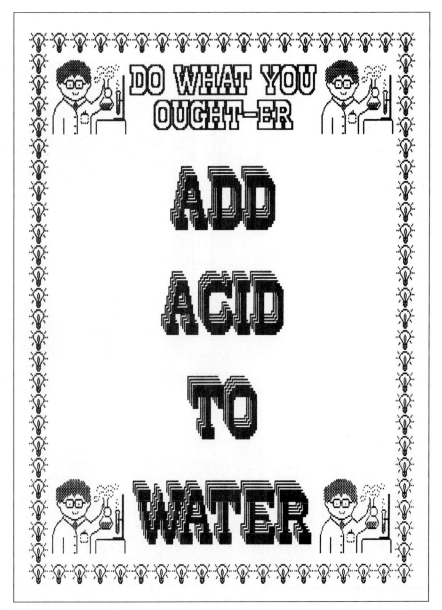

(a)

Figure 7.6 continued. A crossword puzzle on meteorology (screen b) prepared with CROSS-WORD MAGIC.

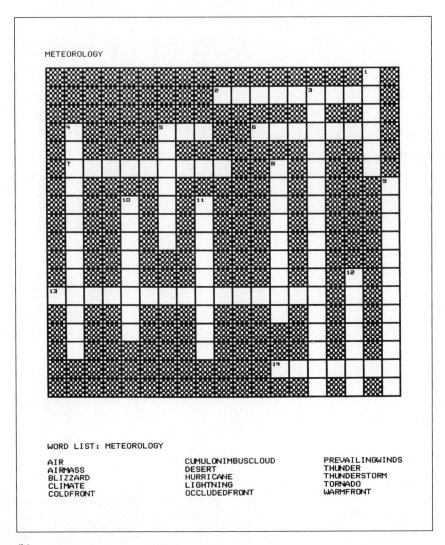

(b)

SUMMARY

This chapter described computerized tools and basic strategies to increase the productivity of teachers and students in the science classroom. Word processing, spreadsheet, and database management programs are useful tools that can fit into the science and many other curriculum areas. At all grade levels, these are powerful tools to release teachers and students from routine, time-consuming, peripheral activities, allowing attention to be

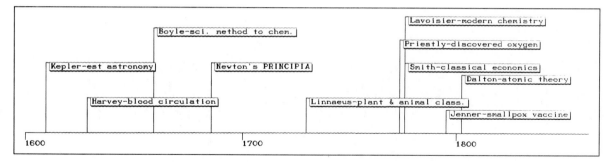

Figure 7.7 A flowchart that shows important scientists and dates of the eighteenth century, which were entered and printed with TIMELINER.

focused on the real goals of science education. In addition, the computer's power to help collect and analyze data offers a valuable tool for testing hypotheses and conducting scientific research. Finally, test-generating and graphics programs can serve as tools for specific tasks in the science classroom. A very specific tool application of the computer, the microcomputer-based laboratory, is the subject of Chapter 8.

REFERENCES

Dinkheller, A., J. Gaffney, and E. L. Vockell. *The Computer in the Mathematics Curriculum.* Watsonville, Calif.: Mitchell, 1989.

Pogge, A. F., and V. N. Lunetta. "Spreadsheets Answer What if...?" *The Science Teacher* 54(8) (November 1987): 46-49.

Schwartz, E., and E. L. Vockell. *The Computer in the English Curriculum.* Watsonville, Calif.: Mitchell, 1989.

Vockell, E. L., and J. Hall, "Computerized Test Construction." *Clearing House* 62 (1988): 113–123.

Vockell, E. L., and E. Schwartz. *The Computer in the Classroom.* Watsonville, Calif.: Mitchell, 1988.

CHAPTER *8*

MICRO-COMPUTER-BASED LABORATORIES

THE DICTIONARY DEFINES THE laboratory as both a place equipped for experimental study and a period of time set aside for laboratory work. Because the laboratory is the place where scientists do a lot of their work, it seems logical that the laboratory ought to be the center of any good science program. In addition, most science programs list as major goals that students understand and use the processes of science in solving problems and making decisions and that they understand the nature of the scientific enterprise. These goals imply the use of the laboratory.

However, as we have suggested earlier, many teachers are unable or unwilling to offer laboratory investigations in which the student is an active seeker of data and analyzer of results. Where laboratory experiments do exist, hypothesis testing has often given way to reaffirmation of concepts already presented by the teacher or memorized by the students. This is particularly unfortunate for students whose learning styles or personal needs demand that they interact with concrete ideas or with other students while learning. The programs discussed in previous chapters can help overcome these problems. This chapter focuses on the microcomputer-based laboratory (MBL) as an aid to collecting and tabulating data and enabling students to become actively involved in the process of science.

THE IMPACT OF THE MICRO-COMPUTER-BASED LABORATORY

The use of the microcomputer as a tool in the science classroom can have a significant impact on the content of the science course; but, equally importantly, it can change the teaching and learning strategies employed. It can assist the teacher in providing opportunities for the students to learn and practice higher-order thinking skills.

Robert Tinker, director of the Technical Education Research Centers (TERC), states:

> Use the computer as a laboratory instrument. Make it into a tool that allows students to quantify the world about them. Give the students a fantastically powerful instrument that no science teacher could have dreamed of having in class only a few years ago—an instrument that can measure force, light, pressure, temperature, heart rate, acceleration, response time, brain waves, muscle signals and many, many other phenomena in the world about us. . . . Give students these tools and you will see a (pardon the expression) revolution in science education—a true embodiment of Piaget's notion that children learn best by discovering and creating the world for themselves (1984, pp. 24–26).

The most important thinking skills practiced in the science laboratory should be problem solving and hypothesis testing. The process of designing, hypothesizing results, asking "what if" questions, and carrying to completion one's own experiment provides an environment for the maximum learning of scientific problem-solving skills. Equipped with simple sensors and software, the microcomputer becomes a laboratory instrument or an MBL. The MBL has been shown to be one of the most powerful strategies for teaching science at the secondary level (Wise, 1989).

Using MBLs helps both teachers and students be active participants in the science process. Teachers can move away from the text toward laboratory experiments that may be more directly related to the world of the student outside the classroom. As a result, teachers lecture less, and students become involved in the active quest for information. This will necessarily cause a change in the classroom environment, including the objectives taught (more process-oriented skills), the teacher's role (facilitator instead of information giver), and the evaluation of students (problem-solving tasks rather than factual memorization questions).

ADVANTAGES OF MICRO-COMPUTER-BASED LABORATORIES

A thoughtfully designed laboratory-based curriculum can benefit greatly from MBLs because the MBL is often especially well suited for implementation in the learning cycle approach. They are particularly useful in the exploration phase, enabling students to come to grips with the nature of the system under consideration—observing it, asking "what if" questions, and formulating new questions based on preliminary data. In the concept development phase, the teacher can use the MBL in a demonstration mode to clarify ideas and focus attention on key concepts by using data actively collected by students. In the concept application phase, students can use the MBL to efficiently conduct a number of experiments to broaden and generalize the concept they have developed.

Another advantage of MBLs is that they reinforce several learning modalities (Mokros, 1986). Students must manipulate the laboratory materials, stimulating their kinesthetic learning modality. They can watch their science experiment changing or watch a graph on the computer screen and activate their visual modality. With many MBLs, sound stimulates the auditory modality. In a science experiment, therefore, students may listen to sounds while watching a graph of the pitch rising or falling and begin to strengthen modalities that they do not automatically use. Learners are more successful if they can use all modalities and move from one to the other easily (Woerner and Stonehouse, 1988).

In addition, MBLs link the concrete experience with more abstract and symbolic representations such as graphs and equations. Students can attach a probe to measure heart rate and watch the graph on the screen. Every move and every beat can be seen and felt as they appear on the graph. Thus, students know what the graph represents.

Using MBLs provides an opportunity for students to gather *and* analyze data in an active manner, much like professional scientists. The laboratory becomes a place where they spend time in active mental activity, not in routinely collecting and recording data. In traditional laboratories, students spend most of their time with menial activities like writing the temperature or the value of some other dependent variable every twenty seconds, with the result that they often have very little time left at the end of the laboratory for data analysis. In addition, the traditional lab often permits the collection of only one set of data, and students are often sent home to analyze that data as homework, without the benefit of dialogue with other students or with the teacher. This situation discourages the teacher and the student from taking seriously either the concept development or the problem-solving aspect of science laboratories.

A good example of the use of MBL to overcome problems with traditional laboratories is the melting point experiment. In this laboratory, which is conducted in almost all secondary chemistry classrooms, students use a thermometer to determine the melting point of a substance, such as mephthalene. As the experiment progresses, students watch the clock and read the thermometer, which is placed in the substance contained in a water bath. Students must take the temperature of the substance every thirty seconds, until it melts. Too often, students are so intent on watching the time or reading the thermometer that they do not observe what is happening during the experiment. They very seldom see the melting occur. Their resulting data are incomplete or inaccurate. Students then take their faulty data home to construct graphs and determine outcomes. Their experiment is discussed the next class day. The discussion is typically not helpful to students who have spent very little time carefully observing and analyzing the phenonema under consideration.

Let's look at this same melting point laboratory activity in an MBL environment. By using a computer interfaced to a temperature probe placed in the substance, students can watch the temperature change in real time on a graph, plotted as it occurs. They will be free to observe the substance, so that an accurate determination of the melting point can be made. As the experiment progresses, students will have the opportunity to intervene and change any condition that is not correct or to run a slightly modified experiment if the results dictate. Students can also discuss the reasons for their results while they are being obtained. Further investigation can be made if discrepancies are found in the data or if the results cannot be explained.

With MBLs, the computer collects data in real time, and the student watches the experiment in progress. This encourages active concept development and the application of process skills. Science learning becomes more concrete and intimately connected to observable phenomena. Brasell (1987) observes that this real-time graphing can be the key to the understanding and interpretation of distance and velocity graphs in physics. Neither producing graphs later by hand nor being shown a graph after the experiment is over can produce the same learning as examining the graph as it evolves in real time. In comparing computer-based approaches in science teaching to traditional instruction, Wise (1989) finds that computers have an overall positive effect on learning but that the strongest effect for all types of computer-based learning was found for MBLs.

Note that in our examples we have shown how the computer can save time by serving as a tool to perform an onerous and distracting task. We *assume* that students will reinvest this time in more active and productive learning. If this assumption is false, then there will be no enhancement of academic learning time, and the overall outcome will be that the MBL will provide a more expensive path to ineffective learning. To be effective, science activities accomplished with an MBL must be meaningful and engaging to students. Moreover, the teacher must use a learning cycle approach to help students focus on worthwhile aspects of the experiment. Only then will the use of MBLs lead to the development of both scientific problem-solving skills and accurate science concepts.

COMPUTER INTERFACES

The Apple II family of computers provides a simple and inexpensive method of interfacing laboratory experiments to the computer. Teachers or students with technical expertise can construct a simple joystick port input device, which allows for many different kinds of input devices to be attached. This input device measures a difference in resistance across two pins. Various devices, called transducers, can be connected across these two pins, which produce variation in the resistance as specific external conditions change. A software program can be written to record these resistance changes and to convert them to the desired units. Thermistors can measure temperature. Resistors can measure the resistance of a circuit in ohms. Phototransistors can measure the period of a pendulum, acceleration, or changes in light intensity. The attachment of a 100-K ohm potentiometer allows students to measure the height of a rocket's flight, the period of a pendulum, or the speed of a car on the street. Similar devices can be constructed for other brands of computers.

With an understanding of the connective devices and a knowledge of the principles described in the preceding paragraph, science teachers and students with a little engineering ability and creativity can easily construct their own interface devices to measure scientific phenomena. It is not necessary for teachers to build and calibrate their own devices, however; many companies have produced manuals and laboratory interfacing devices that are self-explanatory and require very little technological expertise. Furthermore, the NSTA Honors Seminars, the American Association of Physics Teachers, Project SERAPHIM, and TERC have trained science teachers in this new technology (Layman et al., 1985; Woerner, 1986). These workshops have shown that inexpensive devices and public domain software can be used to provide tools for many good science experiments. These modules and manuals explain how to build your own interface devices. Project SERAPHIM, an NSF-sponsored clearinghouse for instructional microcomputer information in chemistry, has laboratory modules for making adapter boxes and interfacing devices for measuring temperature, heats of reaction, and photochromic kinetics. The AAPT manual by Layman and co-workers (1985) provides a disk that contains programs needed to calibrate and utilize the interfacing devices produced and describes physics applications.

Science teachers participating in summer institutes in northwest Indiana helped write a teacher's manual for *The Interfacing of Microcomputers in the Secondary Science Curriculum Through the Apple Game Port* (Woerner, 1986). This manual (Figure 8.1) contains the directions for building the simple, inexpensive interfacing connectors and includes twenty-six experiments in biology, chemistry, physics, and Earth science, which use the computer as

Figure 8.1 A page from a teacher's manual showing how to build a model rocket–tracking device. The rocket tracker was designed by Mike Slootmaker, a middle school teacher in Fremont, Michigan. The design appeared in *Science Scope* 8(3) (February 1985): 6–8.

EXERCISE 5: MEASURING HORIZONTAL AND VERTICAL DISTANCES

A potentiometer behaves exactly as the paddle inputs did. By moving the potentiometer knob, a variable resistance is read into the computer using the PDL(Ø) and PDL(1) commands. The resistance needs to have an effective range of Ø to about 15Ø,ØØØ ohms.

ROCKET TRACKER*

To build the rocket tracker designed by Mike Slootmaker, you need a 1ØØ K ohm <u>linear</u> potentiometer (Radio Shack part #271-Ø92. Cost $1.Ø9) and your 16 pin or 9 pin interface connector. We will connect our tracker to the paddle Ø wires.

Obtain a potentiometer and cut 1/4 inch threads onto the handle. Using about 3.3 meters of flexible, multistrand wire, solder the wire to the leads on the potentiometer as shown here.

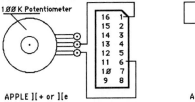

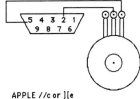

APPLE][+ or][e   APPLE //c or][e

Obtain a metal shelf brace and enlarge the screw holes. One of these holes must allow the potentiometer to fit snugly against the brace. The other hole must fit the screw on the top of a photographic tripod and will be fastened on with a wing nut

to paddle Ø →

The sighting device itself will be a 6Ø cm piece of rigid plastic pipe. We will use 1/2 in. Cresline PVC plastic water pipe A 1/4 in hole should be drilled through the plastic about mid length down the pipe. Attach a nut as far down onto the potentiometer as possible. Slide the plastic pipe on through the hole and fasten it tightly against the nut with a wing nut.

When you move the sighting device from horizontal to vertical, the potentiometer knob should move with the sight.

* Rocket Tracker designed by Mr. Mike Slootmaker, Science Teacher, Fremont Middle School, Fremont, Michigan. "A Computer-Interfaced Rocket Tracking System" by Mike Slootmaker. <u>Science Scope</u>. 8:3, p.6-8.

33

the tool to gather the data. The accompanying disk contains calibration programs for the various transducers.

CLASSROOM USES FOR MICRO-COMPUTER-BASED LABORATORIES

Most interfacing devices can be used in a one-computer classroom as a demonstration or as a whole-class experiment. Strategies for large-group and small-group use of the computer are described in Chapter 9. In many cases, however, the best strategy is to let small groups participate in the hands-on/minds-on activity at their own workstations. Most experiments can be performed safely around computers with a few precautions.

In Walnut Creek, California, an eighth-grade science class has been participating in a project called "Computers as Lab Partners" (Linn, 1987). Using Apple II computers, probes, and software developed by TERC, students learn about heat and temperature in detail, about how scientists reason, and about how to design and interpret graphs. The students easily handle concepts about calibration, scales for graphs, and other problems. They also construct their own concepts based on the results of experiments.

Physics students in Florida and sixth-grade and college students in Boston participated in a unit on motion, using computerized motion detectors developed at TERC. In Florida, physics students have used the unit to study velocity/time and distance/time graphs (Brasell, 1987). Students who saw the graphs forming as the experiment progressed (in real time) performed better on the unit posttest than did students who saw the entire graph printed just after the experiment was completed.

In northwest Indiana, fourth graders are studying their unit on heat and temperature, supplemented with experiments that are done using thermistors and TEMPERATURE EXPERIMENTS (Hartley). These fourth graders have been able to handle the equipment—and warm and cold water—at the computer. They have learned new methods of measuring temperature and have learned intuitively what the graphs on the screen represent. The hands-on experiences with conduction, convection, and radiation were enhanced by letting the computer gather the data from a set of experiments, which lasted for more than one hour. Doing these same experiments without the computer would probably have taken about three hours, and it is extremely unlikely that the more routine activities would have held the students' interest for such a prolonged period of time.

COMMERCIALLY AVAILABLE MICRO-COMPUTER-BASED LABORATORIES

TEMPERATURE EXPERIMENTS (Figure 8.2) contains a well-constructed interface with two thermistors for temperature studies. The program is useful for working with both younger and older learners. The program contains a tutorial designed for upper elementary and middle school students on how to read a mercury-in-glass thermometer. The package includes very sturdy thermistors, which provide good results in the experiments, if the device is properly calibrated. Directions for calibration are clear and require the use of an accurate thermometer. Readings may be done in Fahrenheit or Celsius. Students have the option to observe temperature changes from one or both thermistors by viewing them on a CRT thermometer or on a graph plotted with digital readings.

Students watch the real-time graph being formed as the experiment progresses. Graphics can be saved and recalled for later analysis and discussion. Two graphs can be superimposed on each other, and graphs can be printed. The user can change the scales of the graph before data collection and enter text describing the experiment being conducted. The resulting printout contains a short lab report as well as the graph. The user can pick various time scales from two minutes up to seventy-two hours.

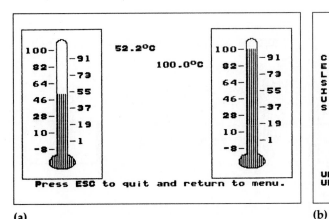

(a)

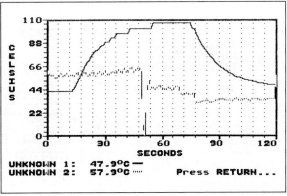

(b)

Figure 8.2 Screen (a) from TEMPERATURE EXPERIMENTS shows two thermometers, which represent two thermistors, in water of different temperatures; screen (b) shows a real-time graph which is the result of students moving one thermistor between two beakers of water. The second thermistor is on the desk.

This longest time period is useful for placing a thermistor outdoors to record temperature changes over a weekend.

PLAYING WITH SCIENCE: TEMPERATURE (Sunburst) is another excellent program for measurement of temperatures with thermistors (Figure 8.3). This program comes equipped with three workable, color-coded thermistors. The extensive manual contains teacher directions and student blackline masters for many kinds of experiments. Hidden utilities allow the teacher to select printer options, set the temperature scale to Celsius or Fahrenheit, and to calibrate the probes. A help option is available at all times. The user can decide how many thermistors to use, the range of temperatures to be recorded, and the amount of time to record data. Students can change the type of graph or digital readout and can adjust the graphs' scale even after conducting the experiment.

Broderbund has designed a series of SCIENCE TOOLKITs for all types of science experiments (Figure 8.4). The MASTER MODULE contains an interface box, a thermistor to measure temperature, and a photocell to measure light intensity. The MASTER MODULE can be used as an accurate stopwatch or a strip chart recorder. Several experiments are suggested in the accompanying manual.

Three additional modules can be attached to the MASTER MODULE. In the BODY LAB module, the user builds a cardboard spirometer into

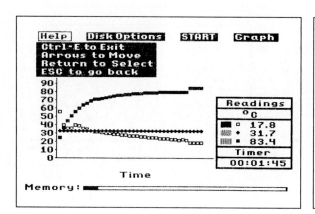

(a)

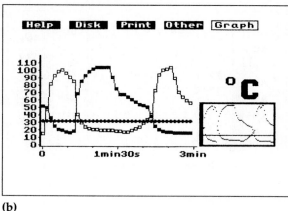

(b)

Figure 8.3 Screen (a) shows a real-time graph of three thermistors showing the help option from PLAYING WITH SCIENCE: TEMPERATURE. Screen (b) shows a graph that is the result of the computer analyzing and condensing the real-time graph.

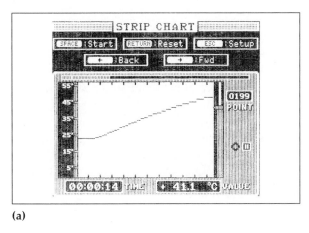

(a)

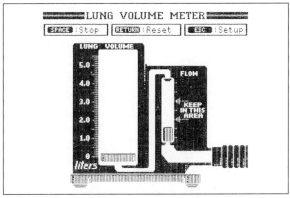

(b)

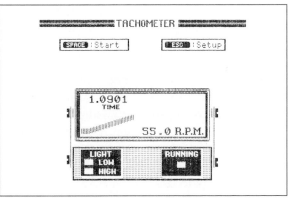

(c)

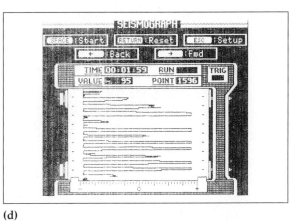

(d)

Figure 8.4 Several screens from SCIENCE TOOLKIT. Screen (a) shows a strip chart recording of water being heated in a beaker. Screen (b) shows the lung volume meter from the BODY LAB module. Screen (c) shows a tachometer measuring the period of a pendulum. Screen (d) shows a seismograph from the EARTHQUAKE LAB module.

which the light probe from the MASTER MODULE is placed. As the student blows into the spirometer through a straw, the vanes inside the spirometer spin, alternately cutting off light to the light probe and then allowing in more light. This change in light intensity and the circumference of the spirometer are used to calculate the student's lung volume. With the "Heart Rate Meter," the user takes a pulse manually for ten beats and then enters that number. The computer calculates the heart rate and prints a graph based on a constant rate. There is also a response timer on this disk. This module contains printing drivers and information, allowing graphs from all Broderbund modules to be printed.

The EARTHQUAKE LAB allows the teacher or student to build a "seismoscope," which is a cardboard device similar to a pendulum seismoscope used in early earthquake detection. As the seismoscope is shaken, a lever arm with a paper tab moves up and down in front of the light probe, and this is converted into a graph on the screen, which resembles a seismograph. The module on SPEED AND MOTION uses two light probes and a balloon-powered car to measure speed and acceleration. The manual suggests other experiments involving the measurement of speed on the tachometer or on the speedometer. In addition, this module can be used to carry out pendulum experiments.

David Vernier (1986) has been building and distributing plans and kits for the "build-it-yourself" interfacing enthusiast for several years. His area of interest is physics, and most of his devices have been developed in that area. *Build a Better Mousetrap* includes directions for a photogate timer, reaction time, microphone/amplifier, humidity meter, IC temperature probe, voltage monitor, thermocouple, pH meter, strain gauge, optically isolated switches, computer-controlled car, stepper motor, and "a better mousetrap." His temperature plotter, voltage plotter, frequency counter, and photogate systems come complete with instructions, parts, and disks with software to calibrate and run the instruments. He will also build the interfaces for you and ship them ready-to-use.

The Vernier PRECISION TIMER II package uses photogates and the computer as an accurate timing device. Activities to determine the speed of a rolling or falling object, pendulum studies, oscillating objects, air track experiments, and many other studies involving the speed and acceleration of moving objects and times can be done with this kit. Both Apple and IBM versions are available. The TEMPERATURE PLOTTER III connects either two or four thermistors to the computer and has various graphing options and temperature alarms.

FREQUENCY METER III (Figure 8.5) monitors and analyzes audio-range frequencies. It may be used to determine the frequency of any signal that can be converted to voltage sine or square waves. The program allows information to be read by the computer through an interface device containing a microphone/amplifier. Whistles, tuning forks, and other devices can be used to produce tones. There is both real-time graphing as well as a "strip recorder" which displays a permanent record of frequency

Figure 8.5 Screen from FREQUENCY METER III showing an audio frequency plotted over time.

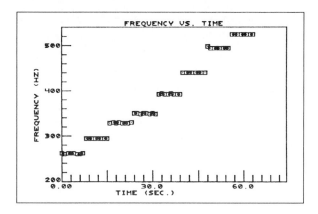

readings. Frequencies can be displayed digitally as well as graphically, and the numbers can be seen in large screen or regular format.

Project SERAPHIM, an NSF-sponsored clearinghouse described in Chapter 5, offers both interfacing information and CAI programs in chemistry education. The Project's laboratory modules include computer disks and instructional objectives. The teacher, however, must obtain and assemble the interface materials. The programs cover such topics as temperature measurement, heats of reaction, and photochromic kinetics.

Several programs produced by TERC became commercial MBL packages originally marketed by HRM and now marketed by Queue. EXPERIMENTS IN SCIENCE, which includes disks, probes, sensors, interface box, and cables, allows students to conduct experiments in biology, physiology, chemistry, physics, and Earth science. The CARDIOVASCULAR FITNESS LAB uses a light probe attached to the earlobe or a finger to measure "heart rate." In actuality, small differences in the amount of light conducted occur as the blood is pumped into the earlobe or finger as compared to when the heart is at rest. This change in light intensity is converted into a number representing the heart rate, and a graph shows the heart rate as it may change. Sound may be obtained to simulate the heart beat.

Cross Educational Software also produces interfaces and probes for many different science experiments, although the user must assemble all the instruments. Thornton Associates produces an interface called the SPI SYSTEM (Sensor-Processor Interface). SPI SYSTEM adapters are available for voltage setups, temperature probes, microphone interfaces, light/ phototransistors, Geiger counters, electrocardiographs, and respiration monitor attachments. Experiments in most science areas can be conducted using this system. The SPI SYSTEM interface comes with software for the Apple II or for MS-DOS computers. Thornton also has a MACSCOPE interface for the Macintosh computer. LEAP also produces an interface marketed by science supply companies, which can be used with many different types of sensors, including photovoltaic probes, high-temperature thermistors, Kjeldahl probes, conductivity probes, pH probes, and pressure gauges.

The SCI-LAB system (Sargent-Welch) is another interface kit offering many experiments for biology, chemistry, and physics. Probes in this kit are connected through a common interface box hooked to an Apple. This package comes with a "key," which is a microcomputer chip that must be placed inside the computer in the sixteen-pin game input/output port. In addition, an analog-to-digital converter board must be placed in one of the Apple's slots. A Grappler+ interface card is necessary for the high-resolution graphs that the programs produce. The transducers available are temperature, force, electrocardiograph sensors, dynamometer, respiration probe, pH electrode, photogate, a computer-controlled car, and a bubble detector. Each experiment described in the kit comes with its own disk and a short explanation. Chemistry experiments include heat of fusion, indicators, pH, titration of a strong acid, and energy and phase changes. Biology experiments include the electrical activity of the heart, skeletal muscle activity, evaluation of physical conditions, mechanics of respiration, rate of photosynthesis, and rate of yeast fermentation. In physics, experiments involving temperature, forces, velocity, and acceleration can be conducted.

The INTERFACING COLORIMETRY PROGRAM (Kemtec) interfaces to any colorimeter but is designed to run most easily on that company's own light-detecting instrument plugged into the Apple joystick port. Standard colorimetric experiments can be conducted using this kit, such as determining the equilibrium constant for FeSCN. Enzyme kinetics and toxicology experiments are examples of the advanced work that can be done with this kit.

pH PLOT (CDL) has an interface and a pH electrode, but data can also be entered from the keyboard. The software allows the production of a table or a graph from titrations of acids and bases and allows multiple plots on the screen. If the interface is being used, the keyboard can be shut down so students do not "fudge" the data, an important feature in this program.

CHEMPAC (E & L Instruments) is an integrated teaching system for chemistry available for Apple, Commodore, and IBM. This package is a laboratory program that uses the microcomputer as an integral part of the course. The kit contains a sophisticated interface, a suitcase of probes and equipment, and teacher and student manuals. The teacher's manual describes many of the usual types of experiments conducted in high school chemistry. More important, it describes how the computer can help students carry out the activities most efficiently. Each laboratory station has the capabilities of a pH meter, calorimeter, colorimeter, pressure sensor, EMF cell, a thermometer, and other instruments.

The MICROCOMPUTER-BASED LABORATORY PROJECT, designed by TERC and available from Queue, consists of three modules that provide an inquiry environment, encouraging middle school and older stuc'ents to design their own experiments. THE MBL PROJECT: HEAT AND TEMPERATURE module comes with a "heat pulser" that can be placed in a beaker to administer even pulses of heat. This enables students to investigate temperature changes as additional pulses of heat energy are added. The primary emphasis is on real-time graphing, with the student designing and redesigning the experiment. Included with the software are activities challenging students to reproduce a particular graph.

With THE MBL PROJECT: SOUND module, students use a microphone probe to measure sound and observe graphs representing sound waves. Although sound is a very complex and abstract physical phenomenon, students can use this MBL to work with concepts of amplitude, frequency, wavelength, and wave shapes to begin to develop a qualitative understanding of the differences between sounds.

THE MBL PROJECT: MOTION module enables students to use a range detector that sends out brief sound waves and records their return after they have been reflected from a distant object. The computer automatically generates graphs that show distance versus time, velocity versus time, or acceleration versus time. Besides printing graphs, students can simultaneously display two of these graphs on the screen to compare designated variables. By observing the graph drawn in real time, students get a concrete impression of the abstract factors involved in motion.

BIOFEEDBACK MICROLAB (Queue) uses thermistors, a light probe, and probes to measure muscle tension (EMG) and electrodermal activity (EDA) to investigate the nature of electrophysiology. The manual provides suggestions for measuring heart rate, skin temperatures, EMG, and EDA and then investigates what factors affect these readings.

SUMMARY

The computer can become an integral part of the science classroom by automatically collecting, recording, tabulating, and presenting data for science experiments. The number and types of microcomputer-based laboratory devices are likely to rapidly proliferate in coming years. It will certainly become easily possible in the microcomputer-based laboratory to measure phenomena which could never be introduced into the traditional classroom. By using microcomputer-based laboratories wisely and effectively, teachers can help their students focus their energies more efficiently on the important goals of science education.

REFERENCES

Brasell, H. "The Effect of Real Time Laboratory Graphing on Learning Graphic Representations of Distance and Velocity." *Journal of Research in Science Teaching* 24 (1987): 385–395.

Layman, J., M. DeJong, and J. Nelson. *Teacher Tutorial for AAPT Microcomputer Workshop on Laboratory Interfacing Experiments Using the Apple II Microcomputer Game Port*. Washington, D.C.: American Association of Physics Teachers, 1985.

Linn, M. "An Apple a Day." *Science and Children* 25 (November/December 1987): 15–18.

Mokros, J. R. "The Impact of Microcomputer-Based Labs on Children's Graphing." *TERC Technical Report* (February 1986): 1–8.

Tinker, R. "The Decline and Fall of the High School Science Lab." *Electronic Learning* 3(5) (1984): 24–26.

Vernier, D. *How to Build a Better Mousetrap and 13 Other Science Projects Using the Apple II*. Portland, Or.: Vernier Software, 1986.

Wise, K. "The Effects of Using Computing Technologies in Science Instruction: A Synthesis of Classroom-Based Research." In J. D. Ellis (Ed.), *Information Technology and Science Education*. Columbus, Ohio: ERIC, 1989.

Woerner, J. J. *The Interfacing of Microcomputers in the Secondary Science Curriculum Through the Apple Game Port*. Gary: Indiana University Northwest, 1986.

Woerner, J. J., and H. B. Stonehouse. "The Use of the Neuro-Linguistic Programming Model for Learning Success." *School Science and Mathematics* 88 (1988): 516–524.

CHAPTER 9

SELECTING AND USING COMPUTERS AND SOFTWARE

SOME PRACTICAL PROBLEMS MAY arise when teachers try to introduce the computer into their science classrooms. This chapter identifies some of these problems and suggests solutions.

SOFTWARE SELECTION

The key to selecting software is choosing software that actually fits the needs of your curriculum and students. Resist the temptation to use a piece of software just because it's "a great program." A program is a good program for your students if it helps them attain the objectives they need to attain.

Vockell and Schwartz (1988) provide detailed guidelines for selecting drills, tutorials, and simulations. This section summarizes just a few guidelines of particular importance in selecting science software:

1. Select software that is compatible with your overall approach to science instruction or that supplements your major approach.
2. Select software that teaches skills that are not easily taught through other media. However, be aware that there can be problems with teaching aspects of science as isolated skills. Skills taught in isolation must be integrated with the "big picture."
3. Select software that compensates for weaknesses in your curriculum or in teaching style. If your textbook is heavy on factual treatment of concepts, select programs that emphasize process skills. If you have a rapid-fire mode of presentation, look for some programs that teach the same skills at a slower pace.
4. Select programs that promote cooperative learning (which is discussed in detail later in this chapter).
5. When selecting programs for higher-level objectives, look for programs that can be adapted either for a specific portion of the learning cycle or throughout the entire cycle. It is especially important to look for software suitable for large groups during the learning phase—especially during term introduction—and for small groups during the practice or application phase of instruction.
6. Select software that lets you adopt new approaches that would not be possible or feasible without computers. The National Science Teachers Association (NSTA) has distributed a detailed software evaluation instrument that is useful for assessing the instructional quality, the technical quality, the subject matter, and the appropriateness of software

for science teaching. This instrument is useful for curriculum developers, researchers, and media center personnel who are selecting software for instructional use in science.

It is impossible to select appropriate software without considering the situation in which it will be used. A program that is "good" for one teacher or in one situation may be "terrible" in another. The best way to evaluate the potential usefulness of a piece of software is to perform a task analysis of the instructional activity for which it will be used and then check to see how well the computerized material fits into this instructional unit. For example, imagine that a teacher is planning to teach a unit on weather fronts. The major goal of this unit of instruction is that students will understand the basic concepts *warm front* and *cold front*, so that they can better understand the weather reports on local TV newscasts. It happens that there are two programs entitled WEATHER FRONTS, one by Teach Yourself by Computer and the other by Diversified Educational Enterprises. Before selecting a software package, the teacher should ask, "What do I actually want to *do* with the program? Why do I need a computer program at all?" Let's assume that the answer to this question is that the teacher wants to give a good introduction to the concept by providing examples and nonexamples, encourage discussion to identify critical attributes of the concept, and then have students interact with the computer program to verify that they have understood the concept of weather fronts before going on to the more advanced topic of weather prediction.

In the preceding description, the teacher is actually using the computer as a "surrogate tutor." In selecting software, therefore, it is appropriate to list the tasks that the teacher would ask the tutor to perform, and then see how well the computer performs these same tasks. Table 9.1 lists the tasks that a good tutor would perform for individual students being tutored on weather fronts, if an individual tutor were available. (Note that the same lists of tasks could be used for verifying the understanding of almost any noncomplex scientific concept.)

Figure 9.1 shows several screens from WEATHER FRONTS (Teach Yourself by Computer). If our teacher examined this program, she would discover that it performs Steps 1–3 fairly well. It also performs Step 4; but if Step 4 is performed more than once, the computer simply alternates between two explanations with no increase in clarity and no improved focus on the concept under consideration. The computer does not perform Steps 5–8 at all. It performs Step 9; but because of some weak questions and

Table 9.1 Nine steps typically peformed by a human tutor during the term introduction and application phases of the learning cycle to verify that a student understands the concepts "warm fronts" and "cold fronts." (Note that this task analysis is offered purely as an *example*. We are not suggesting that all science units should follow these steps.)

1. Give a verbal summary (if necessary) of the concept or have the student give a verbal summary of the concept. (This step can often be skipped, and the tutorial sequence can begin with Step 2.)
2. Ask a question about the concept (or part of the concept). A correct answer to this question should demonstrate an understanding of the concept.
3. If the student answers the question correctly, give positive feedback and then ask additional questions, if more are necessary to verify understanding.
4. If the student answers the question incorrectly, provide a verbal restatement of the concept. This should paraphrase the original statement, specifically focusing on the misunderstanding that led to the incorrect answer.
5. Instead of Step 4, offer prompts to stimulate the student to think more carefully and thereby generate the correct answer. This is often a better idea than Step 4, especially if you can focus on the student's misconception.
6. If the student continues to make errors, repeat Steps 4 and 5. If the student seems to be getting closer to the right answer, it may be appropriate to repeat these steps even more often. If the student is making no progress, move on to Step 8.
7. If the student gives the right answer after Step 4 or 5, ask another question to verify that the student really understands the answer to the question. Eventually ask questions that will apply the concept in settings that require generalization of the concept. Repeat Steps 4 through 6 as necessary.
8. If the student shows a persistent misunderstanding of the concept, direct the student back to an earlier phase in the learning cycle, providing concrete experience with the concept under consideration. When the student is ready, return to Step 1 of this tutorial sequence.
9. When the student has answered sufficient questions to demonstrate mastery of the concept, certify that he or she is ready for the next step in the learning sequence.

a lack of branching, it is actually fairly easy for a student who knows very little about weather fronts to be "certified" as understanding the concept.

Is this a good program? For students who have already attained a reasonable understanding of the concepts, this program may work as well as a good individualized tutor because these students would need only Steps 1–3 and 9. They would not need Steps 5–8. However, for students in need of remediation and refocusing, the program would be far less effective. Some students needing help would get this help in Step 4; but the program provides no prompting (Step 5), and many students would not benefit from the rephrasing of concepts supplied by the program. An even more serious problem is that some students who did not understand the concept would get credit for "correct" answers because of faulty questions

```
A COLD FRONT IS FORMED WHEN COLDER AIR
REPLACES A WARMER AIR MASS. THE SLOPE
OF THE FRONT IS VERY STEEP AND SLOPES
TOWARD THE BODY OF THE AIR MASS. SINCE
THE SLOPE IS DRASTIC AND OPPOSITE TO THE
DIRECTION IT MOVES, THE FRONTAL ZONE IS
NARROW. THE FRONT TRAVELS IN THE
NORTHERN HEMISPHERE FROM WEST TO EAST.
AS A FRONT PASSES OVER AN AREA THERE IS
A RISE IN BAROMETRIC PRESSURE, AN ABRUPT
CHANGE IN TEMPERATURE AND A CHANGE IN
WIND, SPEED, AND DIRECTION.
          PRESS SPACE BAR TO CONTINUE
```

(a)

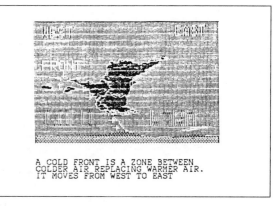

```
A COLD FRONT IS A ZONE BETWEEN
COLDER AIR REPLACING WARMER AIR.
IT MOVES FROM WEST TO EAST
```

(b)

```
              REVIEW QUESTION

THE SLOPE OF A COLD FRONT IS

          A) VERY SHALLOW & BEHIND FRONT
          B) VERY STEEP & FORWARD OF FRONT
          C) VERY STEEP & BEHIND FRONT LINE
TYPE A, B, OR C:
```

(c)

```
A COLD FRONT
      1) IS FORMED BY A COLD AIR MASS
    REPLACING WARMER AIR.
      2) HAS A STEEP SLOPE IN THE OPPOSITE
    DIRECTION TO THE DIRECTION OF TRAVEL
    OF THE AIR MASS PRODUCING A NARROW
    FRONTAL ZONE.
      3) MOVES FROM WEST TO EAST.
AS THE FRONT PASSES A PLACE IT PRODUCES:
      1) A RISE IN BAROMETRIC PRESSURE
      2) DROP IN TEMPERATURE
      3) CHANGE IN WIND DIRECTION & SPEED
            PRESS SPACE BAR TO CONTINUE
```

(d)

Figure 9.1 Several screens from WEATHER FRONTS. Screen (d) shows the remedial information after an error in screen (c).

like that shown in Figure 9.1(d). The decision regarding whether to use this program depends on the students' abilities and understandings and on the teacher's capacity and willingness to intervene and to supply the remediation that is not supplied by the program.

The other WEATHER FRONTS program (Diversified Educational Enterprises) focuses on weather *prediction*. It is an excellent program and may be useful for the next step in our teacher's instructional sequence; but of all the tasks listed in Table 9.1, it performs only Step 1. We repeat: This is a good program, but it is simply not very useful for the goals our teacher is trying to attain.

Besides the steps that appear in a task analysis, there are other considerations in effective software selection. For example, when selecting computer simulations to demonstrate scientific phenomena, the soundness of the model underlying the simulation must certainly be examined. In addition, as we will discuss later in this chapter, if software is to be used for group instruction, the "friendliness" of the user interface must be considered.

Science software continues to proliferate rapidly. Many teachers acquire software by looking for titles that sound like they fit their needs. This is not a wise strategy. Shelves and file cabinets in schools are filled with programs that "sounded good" but have never been used. The task-analysis strategy suggested in the preceding paragraphs may take more time, but it is much more likely to lead to the selection of good software. Choose software that meets your real needs.

SOFTWARE FOR TEACHERS

Besides letting students use the software discussed throughout this book, science teachers can help their students learn more effectively by using the computer themselves. The programs that teachers will run are often substantially different from those the students will use.

In many cases, the most valuable computerized tool available to teachers is the word processor. This tool, which often renders the typewriter obsolete, makes it easier to send personalized notes to children and parents, to write newsletters, to generate and revise tests and handouts, and to perform all the other written tasks that a science teacher must perform. Often teachers can use the same word processing program that their students use; in other cases, teachers will want a more sophisticated program.

Another useful tool for teachers is a gradebook program. Instead of a gradebook program, some teachers prefer to use a database management or spreadsheet program. These programs help teachers keep track of student performance. They are especially useful for keeping inventories and in situations where large numbers of records must be kept, as in mastery learning settings and in highly individualized science programs.

A third useful tool is a graphics package. A specific type of graphics package that is especially useful to science teachers is the poster-generating

Figure 9.2 A poster for a science project generated through the use of PRINT SHOP (Broderbund) and MINIPIX #1 (Beagle Brothers).

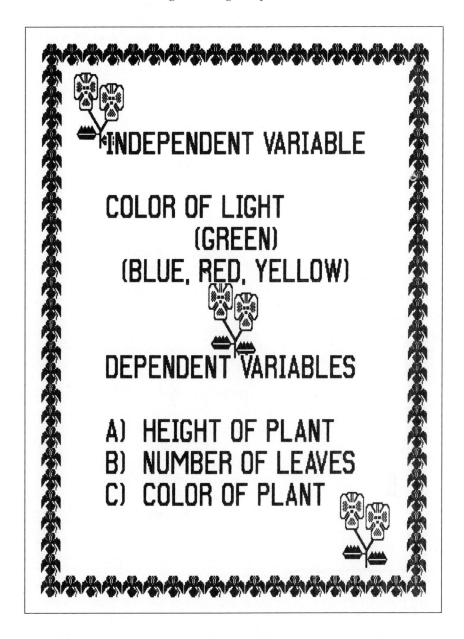

Figure 9.3 A graph of the weight of elementary students on various planets. The information was entered into an APPLEWORKS spreadsheet, and TIMEOUT GRAPH (Beagle Brothers) was used to create the graph.

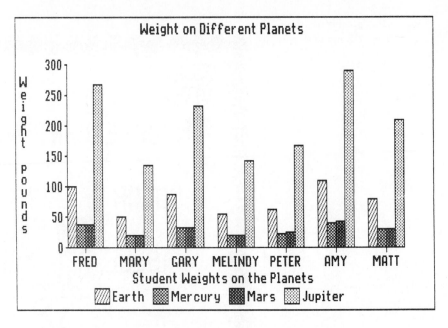

program and the graphing program. For example, PRINT SHOP (Figure 9.2) permits teachers or students to design posters or banners related to science projects. Similarly, graphing programs (Figure 9.3) let students interpret data more easily and submit more professional looking reports. Other programs, such as CERTIFICATE MAKER and AWARD MAKER PLUS (Baudville), let teachers print authentic certificates to recognize significant accomplishments of pupils.

Further details about these and other tools for teachers can be found in Vockell and Schwartz (1988).

TEACHER-DIRECTED USE OF COMPUTERS

Most science software is designed so that students can sit directly at the keyboard and run the program themselves. This is not necessarily a good idea. During successful instruction, students must pass through both a learning phase (exploration and introduction) and a practice phase (application). As the name indicates, during the learning phase students have not yet mastered the concept or skill under consideration; therefore, guidance, corrective feedback, and reinforcement are important. On the other hand, once students have initially mastered the concept or skill, they need to practice it beyond initial mastery, and during this phase feedback or supervision from an outside source is often unnecessary. With factually oriented software and simple skills, there is often sufficient guidance within the computer program itself to let students profit from working alone or in small groups without teacher interaction even during the learning phase. With higher-order programs, however, the software is rarely designed to provide effective guidance, feedback, and reinforcement; therefore, the presence of a competent teacher as a facilitator and coordinator during the learning phase of instruction is crucial. Even excellent software is often trivialized by students trying to guess the right answer rather than working to understand and apply the concept or thinking strategy under consideration.

We are not suggesting that if you encounter a student who is effectively learning a new scientific thinking skill by herself, you should tell her to stop. There is nothing inherently wrong with students running higher-order programs in small groups or even individually. Such groupings may be especially desirable under circumstances such as the following:

- After students possess a sound initial mastery of the skills involved in the program, they may need to pursue additional practice to make the skills part of their automatic, individual repertoires, ready to apply them in appropriate situations.
- After students have initially learned a science skill, there may be four or five good programs that provide additional practice; and it may be better to let students choose one or two of these programs that interest them rather than forcing the whole class to run a single program.
- When cluster grouping is used for gifted students, there may be times when these talented students should work together on a project while the rest of the class pursues tasks that these gifted students have already thoroughly mastered. (The term *cluster grouping* refers to the strategy of putting "clusters" of gifted students in heterogeneous classrooms with a trained teacher who will stimulate them intellectually and encourage them to interact with one another. This is a soundly based alternative to putting them in special classes for gifted students.)
- When slower students need additional practice to master skills for which faster students need no more practice, they can often benefit from working privately or in small groups to practice skills they have initially learned but which have not yet become automatic.

- When specifically structured peer-tutoring programs have been established, in which the tutors and the tutees have well-defined roles and the tutors know how to provide instruction, feedback, and reinforcement, students can benefit from working together at computers. (Just telling students to "work together at the computer" usually does not qualify as structured peer tutoring.)

However, we urge teachers to be cautious and to avoid the overuse of individualistic learning with science programs. Students are likely to waste a great deal of time by using these programs when they cannot guide themselves and when they have no skilled teacher to guide them.

During the learning phase of instruction (exploration and concept development), software designed to teach higher-order thinking skills is best used in an environment where persons with a true understanding of the concept (such as the teacher) can pose problems and ask questions that cause students to stretch their understanding. This can be done as students work in small groups under close teacher supervision during the exploration phase. Another good way for teachers to provide this guidance is to run computer programs with a large group of students, employing either a large monitor or an overhead display device and leading students in a group-interactive use of the software. Students cannot, will not, and do not automatically ask these same questions of themselves or each other. Professional educators with excellent credentials often find this type of teaching challenging; asking students to learn to pose problems, react nonevaluatively, and critically question each other seems unfair. If teaching thinking skills is difficult (and it is), how can we assume that two or three students will teach each other or themselves effectively? Students can (and should) practice together; and sometimes students can help other students understand and master skills, but this is comparatively rare.

Peer tutoring of factual information in science is very effective. Peer modeling of science process skills is also very effective. But peer teaching of process skills is not likely to be effective; the teacher should maintain responsibility for monitoring progress in process skills and prompting their development through effective discourse.

Teachers often assume that it is appropriate for students to go to the computer lab to practice their skills alone after they have introduced the software. Because they can become more actively involved in more problems, it is probably true that students can greatly benefit from *practicing* thinking skills alone or in small groups. But there is a crucial distinction between learning and practicing a skill. To repeat: Our belief is that many teachers make the mistake of sending students off to practice skills that they have not yet learned. This is almost certain to be counterproductive. Thinking is best produced and challenged with discourse. Effective discourse requires knowledge not only of the software and its mechanisms but also of how students learn, how groups interact, and how responses affect process.

Before sending students to work alone or in groups, the teacher should have ascertained that this relatively unsupervised approach is the best way to produce the kind of learning that is desired. In a learning cycle approach, students work in groups with teacher facilitation during the exploration phase, under even more direct teacher guidance during the concept introduction phase, and relatively independently during the practice or concept application phase. Teachers sometimes incorrectly assume that because students are working at computers, they should work independently during the entire learning cycle. This is very often a false assumption that will lead to very ineffective learning.

USING THE COMPUTER WITH LARGE GROUPS

With software that teaches thinking skills, it is often a good idea to work with the whole class or at least a large group during the learning (exploration and concept introduction) phase of instruction. This by no means eliminates the use of the computer. Some programs are easily adapted to large-group instruction. Vockell and van Deusen (1989, Chap. 8) provide effective lesson plans for using the computer to teach higher-order thinking skills to large groups using teacher-facilitated discourse. Because science process skills are examples of higher-order thinking skills, readers of this book would benefit from examining and applying those lesson plans.

An exciting classroom atmosphere often arises when the teacher projects a thinking skills program onto a twenty-five-inch monitor or projection device for the entire class to view. The teacher changes from evaluator to facilitator, from the person who knows answers to the person who asks questions. The computer can be thought of as another teacher who has come into the classroom to assist in teaching the lesson. This "other teacher" poses the problem and provides feedback to the students. The original teacher has the task of helping the students by asking questions and posing problems to challenge them to fully develop their ideas and hypotheses and by responding to them in such a way as to keep them involved in the process.

Chapters 4–6 provided a few examples of abbreviated lesson plans for programs that can easily be adapted to large-screen presentation. For all these programs, the teacher should project the program onto a large screen, so that all students in the group can see it. The group can be either the entire class or a group designated by the teacher. In addition, it is important that the teacher run the program ahead of time and be thoroughly familiar with situations that are likely to arise.

Unfortunately, not all science programs can easily be used in the way demonstrated in these examples. The software must have a simple user interface so that the allotted time is spent thinking, hypothesizing, conjecturing, inferring, analyzing, and testing rather than typing, becoming frustrated, and swearing. The program should allow for collaborative learning so that small groups can attack the problem as a way of collecting input for the whole class. The program should allow for modeling of the thinking process by the teacher. If the program is timed or uses an invariant process, the teacher cannot model the thinking process appropriately.

DISCOVERY LAB (MECC) and DISCOVER: A SCIENCE EXPERIMENT (Sunburst) are both designed to help students develop scientific thinking skills. Each program comes with a booklet that includes a set of objectives, and these objectives look similar (Figure 9.4). Both programs should, at least initially, be used with large groups of students, with a skilled teacher leading the discussion as the program appears on the large screen. DISCOVERY LAB lends itself easily to use by a teacher with a large group of students. The teacher can easily lead a discussion, ask prodding questions, and enter input from the students into the computer to generate immediate responses on the large screen. With DISCOVER, however, this large-group use is almost impossible. For example, to "feed" the creature shown earlier in Figure 6.7(a), the teacher must select MOVE from a special menu, then push

Figure 9.4 Objectives students can achieve using DISCOVERY LAB and DISCOVER: A SCIENCE EXPERIMENT.

DISCOVERY LAB	DISCOVER: A SCIENCE EXPERIMENT
Recognize a simple variable.	Improve note taking and observation.
Design an experiment that controls a variable.	Analyze data collected.
Collect and record simple data from observation.	Formulate hypotheses from data collected.
Provide logical predictions and conclusions from data and information.	Test out formulated hypotheses.
Suggest appropriate experiments to solve simple problems.	Review and reformulate hypotheses based on test outcomes.

arrow keys about twenty times to pick up the food tray and move it into the presence of the hungry alien creature. This is almost impossible to accomplish without distraction while carrying on an effective group discussion. In this case, it is the complexity of the program format (not the complexity of the thought process being learned) that is a stumbling block to effective use of this program for large group instruction.

USING THE COMPUTER WITH SMALL GROUPS

The preceding sections discussed advantages of large-group use of computers under fairly direct teacher supervision during the concept introduction phase of instruction. During the practice phase of instruction, students can use computers more efficiently alone or in small groups, relatively unsupervised by the teacher. While working alone is sometimes desirable, solitary work also has its shortcomings. When students work alone at computers, the following disadvantages are likely to occur: (1) The social isolation involved can create mood states (such as loneliness, boredom, frustration, and fear of failure) that interfere with sustained effort to complete learning tasks; (2) students are denied the opportunity to summarize orally and explain what they are learning; and (3) computers cannot provide social models to be imitated and used for social comparisons (Johnson and Johnson, 1986).

The cooperative learning and peer-tutoring literature suggests that small groups can also overcome these difficulties. Students can work in groups in many different ways at the computer. For example, they can work individualistically and take turns using the computer, they can compete to see who is best, or they can cooperate. Current research suggests that the cooperative approach is usually the best (Slavin, 1983, 1986; Slavin et al., 1985). The key components of effective cooperative learning are positive interdependence, individual accountability, and shared responsibility for one another. This means that the success of the group requires that each person have a role and be accountable for that role and that the individual members be interested in helping one another attain important goals.

Closely related to the concept of cooperative learning is that of peer tutoring. Research on this topic indicates that when students tutor their peers, both the tutor and the tutee benefit from the process (Cohen et al., 1982). Furthermore, rarely in real-life settings do individuals perform or solve problems in isolation. Rather, in today's world, we are required, more than ever before, to interact with others. Therefore, it is often highly desirable, during the practice phase of instruction, to have students work cooperatively in small groups rather than alone.

What this all amounts to is that small groups at the computer are often preferable to individual students at the computer, when working on programs that teach thinking skills, including science. In some cases, large-group (whole-class) presentations may be preferred over solitary work at the computer, especially when new material is being introduced and guidance or feedback from a knowledgeable person (teacher) is essential. But once important skills have been thoroughly learned by some members of the class and at least partially mastered by everyone, then small groups may provide a better use of academic learning time (ALT) than a continued large-group session.

In addition, small-group activities at the computer will support the learning of science and other skills only if the group sessions are structured in such a way as to promote their occurrence. Usually, this means incorporating the direct instruction and learning cycle approaches described in Chapter 2, assigning group activities only after the unit has been appropriately introduced to a large group, ascertaining that group members have actual roles that they understand and objectives that they can meet, and teaching students to interact properly with the program and with one another.

The physical placement of computers can also affect group use. Sometimes computers are set up in such a way that makes it impossible for more than one student to be seated at a computer. We believe that a major "technological innovation" of the 1990s will be the introduction and effective use of a second and third chair in front of each computer.

TRANSFER OF SCIENTIFIC THINKING SKILLS

As Reif (1985) has shown, science skills are of little value unless they can be generalized to novel settings. The most important principle in assuring transfer is to have students focus their attention on what they did that was successful. The teacher needs to ask questions that cause students to label the processes used in their thinking: "How did you figure it out?" Concept development depends on language development. Until students put the concept into clear, unambiguous language and then associate a name or label with it, it is practically impossible to use or transfer the concept to a new environment. Therefore, the teacher should always ask students to reflect on what they have constructed (stating the problem clearly, learning from the apparent rejection of a hypothesis, etc.) and prepare language to explain the concept. If necessary, the teacher should provide prompts to help students realize what they have done and to put their insight into words.

The teacher needs to ask reflective questions at the end of the lesson, prompting students to classify the strategies learned in the lesson and to predict in what other situations they might use those same strategies. For instance, after a lesson using DISCOVERY LAB, the teacher might ask students to recall other times in school that they were asked to think as they did with this program. Thoughtful students might think of numerous applications.

When teaching a content lesson requiring one of the previously taught thinking processes, teachers can get students to look back by asking, "What kind of processes did we use when we solved the problem in . . . ?" By teaching the process in isolation and asking the students to predict when they will use it (anticipation of transfer) and by applying the process to content by asking students to look back at the processes used previously (reflection to engender transfer), the teacher can move into and out of context, transferring the processes as necessary. The teacher should constantly look for opportunities to encourage transfer by using the "remember when . . . ? Now let's . . ." rule: *Remember when* we used careful observation and simplification of a pattern to sequence events in DISCOVERY LAB? *Now let's* see how we can use that same strategy with this experiment."

INTEGRATING SOFTWARE INTO THE CLASSROOM

When deciding when and where to teach a piece of science software, the teacher needs to anticipate where the concept, process, or science skill is going to be applied to content. A science teacher may choose to use DISCOVERY LAB (see Figure 9.4) during a biology unit *after* her students have studied hypothesis testing in earlier units of instruction. In this case, her goal would be to generalize this scientific thinking skill to other settings. Another teacher may use the same program before his students cover hypothesis testing. His goal would be to help his students develop a need for more effective strategies to test hypotheses. Another teacher might use the same program in connection with a completely different topic. There might be considerable overlap among the three applications, but the teachers would focus their goals and discussions differently, depending on the role the program is supposed to play in the curriculum.

It is *not* a good idea to buy and use a program simply because it is considered a good program. A good program is one that helps students use their ALT more effectively to reach an important goal. Always have a goal in mind when you direct your students to use a piece of instructional software.

HARDWARE SELECTION

The key to selecting hardware is identifying your software first and then selecting hardware that will run this software. Currently, most of the best science software runs on Apple II computers. In addition, good software that runs on the Apple is abundant for other areas of the elementary curriculum. These are good reasons to purchase Apple II computers for your science program. However, some programs initially written for the Apple have been converted to run on other computers. Some very effective science programs are designed for other computers, especially IBM and other MS-DOS computers.

Besides computers themselves, certain peripheral devices are important for computerized science instruction. A mouse facilitates movement through a number of simulations, tutorials, and microcomputer-based laboratory (MBL) programs, and the increased efficiency may enhance ALT. Many useful programs require two disk drives, with the program disk in one drive and the data disk in the other. Others require the use of a joystick or similar device.

Programs for the newer versions of the Apple II series, Macintosh, and IBM compatibles are very memory-intensive, so that it is usually necessary to have 1 or more megabytes (M) of RAM memory in order to realistically make use of these programs.

For many programs, it is very useful to have a monitor and printer that will handle high-quality color output, so that visuals displayed or printed by programs such as THE DESERT (Collamore) will appeal to young learners. More advanced ink jet or laser printers will greatly enhance printed graphics output. It would probably be wise not to purchase programs that require high-quality graphic output if either your monitor or printer cannot reasonably handle such requirements.

To participate in the telecommunications projects like KIDNET (National Geographic), it is necessary to have an Apple II GS computer equipped with a modem and compatible networking software. A phone line dedicated to the activity at least for a limited period of time on a regular basis must also be available.

Most MBLs have light, temperature, and pH probes that come with the software. It is important that such kits are designed for the computer you plan to use. A kit designed for an Apple II cannot generally be used with an MS-DOS computer.

Make sure you ascertain peripheral requirements ahead of time, so that you are more likely to purchase software that will run in an educationally profitable fashion on your hardware. If necessary, add the cost of the peripheral device to the cost of the new software that you wish to purchase.

Some apparently convenient and inexpensive programs become impractical when you carefully examine the hardware requirements. For example, a $50 program might require the addition of a $200 device to each of twenty-five computers before you can run the program. If this device serves no other useful purpose and will become outdated within two years, it may be better to select a different piece of software.

Vockell and Schwartz (1988) provide more detailed guidelines for hardware selection.

SUMMARY

Software should be selected in such a way as to help students attain the goals of a school's science curriculum. This chapter described some guidelines for selecting science software. In addition, it described some software designed to help teachers with classroom management tasks. Hardware should be selected to run the designated software most effectively.

When organizing students for using computers, teachers often assume that each student should work at a separate computer. Although individual use of the computer is sometimes a wise use of resources, it is often better to have students work in groups. Especially during the learning phase of instruction, it is often beneficial to have a whole class work with the teacher on a single large screen connected to the computer. This chapter offered guidelines for grouping students at the computer and provided sample lesson plans for large-group instruction in science using computer software.

Finally, this chapter presented strategies for transferring skills learned in computerized science lessons and for integrating these skills throughout the curriculum.

REFERENCES

Cohen, P. A., J. A. Kulik, and C. C. Kulik. "Educational Outcomes of Tutoring: A Meta-Analysis of Findings." *American Educational Research Journal* 19 (1982): 237–248.

Johnson, D. W., and R. T. Johnson. *Learning Together and Alone: Cooperative, Competitive, and Individualistic Learning*. Englewood Cliffs, N.J.: Prentice-Hall, 1986.

Reif, F. "Exploiting Present Opportunities of Computers in Science and Mathematics Education." *Journal of Computers in Mathematics and Science Teaching* 5 (Fall 1985): 15–26.

Slavin, R. E. *Cooperative Learning*. New York: Longman, 1983.

Slavin, R. E. *Educational Psychology: Theory into Practice*. Englewood Cliffs, N.J.: Prentice-Hall, 1986.

Slavin, R. E., S. Sharan, S. Kagan, R. Hertz-Lazarowitz, C. Webb, and R. Schmuck (Eds.). *Learning to Cooperate, Cooperating to Learn*. New York: Plenum, 1985.

Vockell, E. L., and E. Schwartz. *The Computer in the Classroom*. Watsonville, Calif.: Mitchell, 1988.

Vockell, E. L., and R. van Deusen. *The Computer and Higher-Order Thinking Skills*. Watsonville, Calif.: Mitchell, 1989.

CHAPTER 10

INNOVATIVE TECHNOLOGY AND FUTURE TRENDS

ANNE JONES, CHAIR OF the science department at Charles Babbage High School, gets up at 6:30 AM and drinks a cup of coffee. The coffee maker had been turned on by a computer interface device. She then walks into her home office and turns on her microcomputer.

From her home workstation, she uses her modem and communications software to access the high school's twenty-four-hour bulletin board to see if anyone had left messages for her since 9:00 last night. One message from Kris, with the time of 1:00 AM, indicates that he was having trouble completing his chemistry homework assignment on redox equations. Miss Jones types a reply to Kris, indicating that there is a software package, in the library's computer center, that might help him with his problem. She then turns off the computer and returns to her coffee and breakfast.

Upon entering the school at 7:15 AM, she checks the daily schedule and announcements that had been entered on the electronic bulletin board the afternoon before by the principal's secretary. She prints out one copy for each of the science teachers and places the hardcopy in their mailboxes.

Miss Jones takes time to say goodbye to ten students who are going to the Museum of Science and Technology to take part in Project Jason. These students are going to watch Dr. Robert Ballard and his associates as they explore the bottom of the Mediterranean in a minisub, examining old ship wrecks, vents of steam rising from the ocean floor, and life forms associated with those vents. The students had studied the telecommunications procedures and had read about the heated water and the biological studies that are to be conducted. They have prepared questions and are ready to send them to the scientists on site via the telecommunications hook-up at the museum.

The school is running on a special schedule today. The science department is hooking into a telecommunications program involving some top scientists and several of the president's advisors on the scientific aspects of the strategic defense strategy. All the seniors enrolled in any science class are meeting in the instructional media center's auditorium, to take part in the discussion. The students will be able to see and hear all participants in Washington D.C. The seniors in the physics class had submitted three questions via a modem tie-in the week before, and these questions will be answered during the discussion. The discussion is part of an ongoing program for high school students over C-Span, and ten different high schools across the country have submitted questions.

The freshman Earth science classes are conducting a microcomputer-based laboratory (MBL), using several computers connected to temperature probes. They are trying to find out what Earth materials (sand, water, or dark soil) will heat and cool most quickly. A set of Apple IIs on movable carts are ready to be wheeled into the Earth science classroom, and the disks and interfacing devices have all been checked by the teacher the night before. These interfacing devices had been built by the teacher at a special conference last summer.

The biology 1 teacher needs the laserdisc player, the large-screen monitor, and an IBM computer. She has put together a sequence on asexual reproduction, which shows slides of various organisms interspersed with diagrams and explanatory material, and ends with a short "movie" segment showing the diversity of organisms that reproduce asexually.

The chemistry teacher is planning to conduct a laboratory on oxidation/reduction reactions, and she has requested the use of ten Apples on carts for the following day. She plans to run a tutorial program for those students who are still having trouble solving redox equations, to use an electronic spreadsheet program to do calculations, and to use a word processing program to complete lab reports.

The physical science students have been measuring the pH of rain water and local surface water for several weeks. They have kept their results in a computerized database and are ready to transmit their information to State Schools Project, sponsored by the U.S. Office of Education and the Technical Education Research Center (TERC). Their data will be added to the information collected by students in the rest of the forty-eight contiguous states. Students will be able to receive the pooled data and discuss the implications of what has been found. Only a few weeks before, some students in Nebraska found a very low pH in a small lake, and this result was totally unexpected. This information had been unknown to scientists with NSF-sponsored programs investigating the nature of the acid rain problem.

Today the physics teacher will use a laserdisc player interfaced to a Macintosh computer and controlled with a HYPERCARD program to study the collapse of the Tacoma Narrows bridge. The teacher plans to use the software and hardware package to run the "movie" of the collapse in real time. After watching this sequence several times, students can call up individual images and, using split-screen techniques, put two selected images side by side. On the Macintosh screen, there will appear short descriptions and equations, explaining what happened and why. Students will have instant access to "replays" in slow motion and to explanations of observed phenomena and will be able to branch to related segments based on their interests and abilities.

As each of the science teachers arrives in the morning, he or she checks the electronic mail to see if students have encountered problems or if anyone else has contacted them. One teacher has a request from a colleague in Oregon who had seen her presentation at the National Science Teachers Association meeting and wants a copy of the paper she had presented. A former student has dropped a message to another teacher, asking if he would forward a recommendation to a local laboratory for a job as a computer technician. A short message from Ward's Scientific indicates that some items on an order were out of stock and have been back-ordered, and Carolina Biological Supply Company notes that the live amoebas, paramecia, and hydra have been mailed the afternoon before.

During her planning period, one teacher sits down at the IBM computer connected to a CD-ROM player and begins a search for ideas for teaching a unit on nuclear energy. The teacher enters several descriptors and turns on the printer; the computer begins printing a list of resources, most of which can be found in the school library or later that night on a large national database.

A new simulation on predator/prey arrived in the mail the day before, and two biology teachers with a common planning period boot it up to see what it does.

Members of the Earth Science Weather Station Club show up at 8:30 to see what weather information has been collected from their weather station on the school roof. They built the entire station and interfaced it to one of the Apple computers. It runs twenty-four hours a day, and different students monitor the station for two weeks at a time. Each day, students record the high and low temperatures, the barometric pressure changes, and the wind speed and direction; they save the data from the last twenty-four hours and then reset the station to run for the next twenty-four hours. Club members also turn on the Apple II GS that is interfaced with a Microsat System and a dish antenna on the school roof. The GOES satellite will be above the horizon shortly, sending them visual and infrared pictures of the clouds and weather systems. They will then have some pictures for their weather forecast. Students also use the modem to contact Accu-Weather to check their predictions against the one on the national network. They take their information to the media center's television studio to prepare for the morning announcements.

Morning announcements today are designed and announced by the speech department and are transmitted to the large-screen TV in each classroom. General announcements are followed by the weather report from the Earth science class and the sports report from the athletic department. A brief skit advertising the drama department's latest production ends the short broadcast. After the live broadcast, all TV monitors show printed announcements, which repeat at the beginning of each class throughout the day. Several large screens in the cafeteria also carry the announcements continuously. Videotapes are cued up for some classrooms, with the media department coordinating all tapes from a central location.

The science department is to meet during the lunch hour for their weekly electronic discussion with other science departments around the city. Today's discussion is on an updated version of the inventory/ordering software, which they have been using for the last year. In addition, information on the upcoming Science Olympiad and the citywide science fair will be available.

At the end of the day, two science department teachers come to the science department office to access CompuServe using a modem. They enter the science teachers forum and discuss with fifty other teachers from around the country whether dissection of frogs should continue in biology classes. Following a heated discussion, the teachers sign off and dial NSTA's twenty-four-hour bulletin board and download the latest information on earthquakes around the world for their Earth science classes. They also check the latest NASA announcements and then sign off for the day.

• • •

We tried to fool you! It is actually misleading to present this anecdote at the beginning of a chapter on future trends. The technology to do everything described is here *now*. What science teachers can expect in the very near future is more widespread acceptance and implementation of these aspects of technology. In addition, we can expect some new developments in technology that will have an impact on science education.

HYPERTEXT

Occasionally, a new medium encourages entirely new ways of thinking. A relatively new concept called *hypertext* is likely to have this kind of impact on communication and data-retrieval processes. Hypertext is exemplified by the program HYPERCARD (Claris) (Figure 10.1). The basic idea behind hypertext is that a reader is presented with a segment of text or a graphic and then can branch immediately to any other segment as needed. For example, a person reading an instruction booklet on how to use a computer usually reads it from beginning to end. With hypertext, however, the user would read the first screen and then go wherever his or her interests or needs suggested. This might mean going to the second screen, or it might mean going into a more detailed explanation of a term mentioned on the first screen. A reader branching to a screen providing a detailed explana-

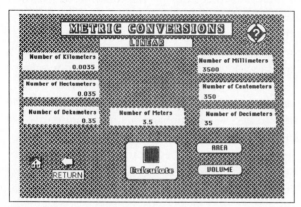

(a)

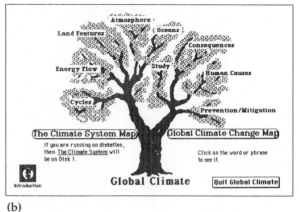

(b)

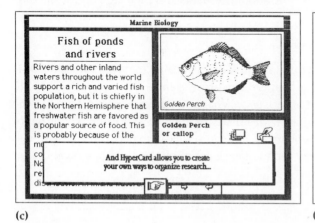

(c)

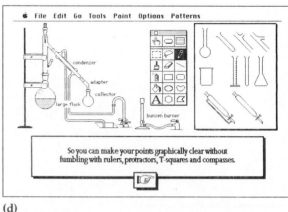

(d)

Figure 10.1 HYPERCARD screens. By clicking in appropriate screen areas, the student has immediate access to information on the designated topic.

tion might go to another screen that furthers this explanation, to a screen that defines or diagrams a term on this explanation page, back to the first screen of the original text, or to the second screen of the original text.

It may have occurred to you that hypertext, when used in this way, is really nothing more than sophisticated, branching programmed instruction. What is unique about programs like HYPERCARD is that they provide an *interface* that makes it easily possible for *both* the designer to program this degree of branching and the user to move through the text. The same strategy that the preceding paragraph applied to an instruction manual could just as easily be applied to a detailed description of an experiment, to a database, to a collection of slides stored on laserdisc, or to almost any topic imaginable. As hypertext becomes more common, it is likely that computer users will become accustomed to the idea of nonlinear thinking. Note, however, that the nonlinear thinking of hypertext is by no means haphazard. Because hypertext requires careful planning and complete sets of information, hypertext programmers and users will learn to incorporate and expect these characteristics in their thinking and presentations. (See the box on pages 234–235.)

One factor that has impeded the introduction of HYPERCARD is that it has been available only for the Macintosh computer. Many districts do not have Macintosh computers or use Macintosh computers only for specialized purposes such as desktop publishing or as file servers for Apple II local area networks. A similar hypertext product called TUTOR-TECH (Techware) has been available for the Apple II computer, but it has not been advertised widely and has had limited impact on the school market. However, the future looks bright for hypertext capability on a variety of computers. Two new comparable programs—LINKWAY (IBM) for MS-DOS computers and HYPERSTUDIO (Roger Wagner Publishing) for the Apple II GS—have been heavily promoted for use in schools. An update of LINKWAY has recently been announced, and Scholastic has announced the development of a hypertext program, HYPERSCREEN, for the Apple II computer.

Applied to science instruction, hypertext will make it possible for teachers and curriculum developers to design learning sequences that include multimedia materials for tutoring students on descriptive concepts in science. The mixing of video, sound, and graphics with text will allow students to vicariously experience phenomena that are either not available locally, too time-consuming, too dangerous, too expensive, or inconvenient to have students experience directly. Even more exciting is the potential for

Common Hypertext Terms

As hypertext becomes more popular, it will become more important to understand terms related to it.

- *Hypertext* refers to the overall strategy of letting users move through structured text in an efficient, orderly manner. HYPERCARD is a specific program that runs hypertext applications on the Macintosh computer. LINKWAY provides a similar service on MS-DOS computers. Programs with hypertext features available for the Apple II series include HYPERSTUDIO, TUTOR-TECH, and HYPERSCREEN.
- A *stack* is the typical hypertext program, which can be imagined as a collection of index cards, each containing systematic information including words, pictures, animated sequences, and sound or a combination of these. These cards are imagined to be arranged in a stack, and the hypertext program lets the user go almost instantly from any point on one card to a related point on another card in the stack. Synonyms include *file*, *story*, and *set*.
- A *card* usually refers to a single screen in the stack. These are sometimes called *pages* or *segments*.
- The *home card* is the starting point in the stack. From this point, the user can branch most efficiently to any other point in the stack.
- Links among cards are referred to as *buttons*, which let the user branch to another card. Some buttons go forward to new text, some go backward to previous text, some ask questions, some play music, and so on. Sometimes the buttons look like actual buttons, are covered by icons that identify functions (e.g., a right arrow to go to the next page), or are invisible (e.g., a screen may include a picture of a halloween screen, with invisible buttons hidden inside the ghost, the witch, the pumpkin, etc.). When the user selects a button, the program branches according to programmed instructions.
- *Clicking* refers to the act of selecting a button, thereby telling the computer where to branch. Clicking usually involves moving a cursor to the button location and pressing a key on the keyboard or a button on a mouse.
- *Hypermedia* refers to the process of integrating several different media with a coordinating hypertext program. For example, a good hypertext program can present text, sound, pictures, and animation—all accessible at the push of a button. COMPTON'S ELECTRONIC ENCYCLOPEDIA is a good example of using hypertext to provide easy access to an encyclopedia supplemented by sound and animation.

continued

continued from p. 234

Interest in hypermedia is expanding rapidly, and new developments and peripherals constantly improve its flexibility. For example, the Apple II video overlay card is a useful circuit board that permits a single video monitor to show both the computer text and the visual presentation from a VCR or videodisc player.

The *HyperStudio Forum* newsletter from the HyperStudio Network, *Stack Central* from A2-Central, and *The Stack Exchange* from Techware offer information about existing stacks for teachers and students. Hypermedia stacks will undoubtedly proliferate in coming years.

using hypertext to simulate various phenomena in a problem-solving setting. For example, the simulation of plant succession in an abandoned field can be done easily and dramatically with hypertext. Video sequences of the entire process can be interfaced with a hypertext program to allow students to change such variables as plant density, time, number of species, and type of environment associated with plant succession and to immediately see the results of their manipulation, along with graphic and text aids that help them interpret those results. Another potential area of benefit is the simulation of chemical reactions. Various combinations of reactants and levels of reactants can be videotaped to provide a database of chemical end products. This database can be used by the developer of the simulation to provide an end product of any reaction or level of reactants that students might choose as they run the simulation.

INTERACTIVE VIDEO TECHNOLOGY

With present technology, teachers often enhance science lessons with films or videotapes. Interactive video technology—the combination of a laser videodisc and a computer—is likely to make these enhancements even more effective. This technology, which is still in its developmental stages, makes it possible to access various combinations of moving pictures, still pictures, text, and sound almost instantaneously. For example, a student might be reading a passage on the computer screen about lunar exploration. The learner might call up a map of the moon. Another keystroke would give an enlargement of an area of particular interest. Noticing several craters, the learner may request a video presentation of meteors impacting the lunar surface and of volcanic activity on the moon. She may wonder why the craters follow certain patterns, and why there are more in some areas than in others. She could have access to all the information on the laserdisc, in any order desired, to try to answer this question. The student could stop the video presentation at any time and then a menu would appear, asking if the computer should move back to an earlier part, skip ahead, change to slow motion, provide a definition, simply pause, or exercise some other option. This is just one example of what can be done with interactive video. Numerous applications are likely to become available as the technology develops and becomes less expensive and more common.

As we indicated earlier, by interfacing laserdisc technology with a program like HYPERCARD, it will also soon become conveniently possible to develop and use first-class databases that include still and moving pictures as well as written text. For example, students will be able to browse through a well-prepared and interesting laserdisc devoted to a specific topic to answer questions and solve problems posed by their teachers or by themselves. Students might watch a video segment that presents a problem, quickly move back to a combination of several still diagrams with accompanying text that provide information to help solve the problem, develop a hypothesis, move to another segment of the laserdisc that shows the results of testing that hypothesis, go back for another look at the problem, go to slow motion or stop action to examine certain aspects more closely, look for additional information on another laserdisc, state a new hypothesis, . . . and continue this process until they have satisfactorily resolved the problem. During this process, of course, students would not be obligated to work solely or even predominantly with the interactive laserdisc. Rather, like other good resources, the laserdisc materials simply serve as a vital resource that is available to provide useful information to individuals, small groups, or entire classes.

Interactive video technology is currently available even without programs like HYPERCARD. However, it is expensive and somewhat difficult to program for specific applications. Within the past year, good discs for science education have become more common, and prices have begun to drop. Our insight is that programs like HYPERCARD will help solve the programming problem. This is because HYPERCARD programmers will simply concentrate on organizing the laserdisc into a useful database from which information can be easily accessed, without consideration for the specific questions users will want to answer. Teachers and students can use the resulting flexible, first-class series of databases to answer questions and solve problems that they consider to be important.

While there is some disagreement to the exact definition of the term *interactive video*, most theorists agree that either a videotape or a videodisc together with some way to program or interact with the video materials are the critical attributes. Daynes (1984) describes three levels of interactive video. In Level 1, there is a very low level of interactivity, with the video player being operated through the manual remote control. Level II involves interaction through a microprocessor, allowing more interaction and limited branching. Level III involves the use of the microprocessor to provide a high degree of branching and much more interaction. At this highest level, the computer and the video machine are working together to bring individualized and interactive opportunities to the learner.

Some science teachers already use science videodiscs. There are many discs available in all areas of science, and more are being developed monthly. The commercial marketing of laserdiscs and laserdisc players is just starting to be big business, and commercial popularity will further stimulate development. A major innovation lies in the area of using the large storage capacity of videodiscs to present visual databases (Sherwood, 1989). Another development is to use the videodisc to present simulations such as PUZZLES OF THE TACOMA NARROWS BRIDGE (Wiley Educational Software), which shows the destruction of the bridge through harmonic oscillation.

Laserdiscs offer options not available through any other media currently used in the classroom. The laserdisc can easily become interactive. With it, the teacher can develop a lesson uniquely designed to teach toward the objectives of the course. Both instruction and testing can be individualized for each student. The use of a videodisc to present real-world problems to students has shown gains in student achievement. Sherwood (1989) used a videodisc of "Raiders of the Lost Ark" to design an interactive computer program that asks students to solve real science questions in some of the segments. For example, at one juncture, Indiana Jones is about to pick up the idol, when he decides that he must place an equal weight in its place. The idol is supposed to be solid gold. Indiana Jones uses a bag of sand and students are asked to decide if there is enough sand to offset the weight of the gold. In this context, students can grasp abstract concepts related to density and appear to be able to transfer the learning to new problems.

Videodiscs currently available for science are listed in Appendix E.

Note that "Indiana Jones" does not appear on this list; so with a little imagination, this list can be greatly expanded.

COMPUTER VIDEO CAPTURE

A recent technological advance has the potential to make interactive video much more versatile. That advance is the video capture card available for both Apple and IBM computers. This card allows the operator to capture and store still or motion video from practically any source—a video camera attached to a microscope, a videodisc, a videotape, a CD-ROM, or a video scanner. The resulting motion or still video image is stored digitally on a high-capacity hard disk drive (80M or greater). Stored video sequences can then be recalled and inserted into hypertext programs or linear sequence programs like STORYBOARD (IBM), without the need to permanently interface with any video device. Video from a variety of sources can be used and integrated with both text and graphics, making an extremely flexible medium for producing multimedia computer-assisted instruction.

MULTIMEDIA APPLICATIONS

Multimedia refers to the integrated combination of the computer with video, audio, text, and other modes of presentation. Like videodisc technology, multimedia applications in the science classroom are expanding rapidly. Both multimedia and videodiscs lend themselves to individualization and to presenting material in many learning modalities. In addition, both recognize that effective learning will use the computer as well as more traditional classroom resources.

The recent development of the VOYAGE OF THE MIMI (Sunburst) materials (Figure 10.2) shows that it is possible to design a package that combines videotapes, printed materials, maps, computer programs, and excellent guidelines in order for the teacher to produce units of instruction that not only excite students about science but also integrate science with skills in language arts, social studies, computer programming, and art. The series includes two television series, microcomputing software, MBL equipment, and printed materials for teachers and students.

The television component consists of thirteen half-hour episodes. In each episode, middle school students are involved in an adventure drama and a documentary/expedition. The activities revolve around the study of humpback whales in the Gulf of Maine aboard the ketch *Mimi*. Various areas

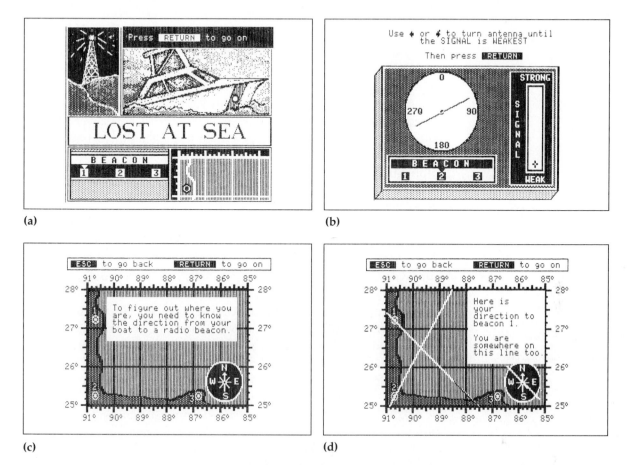

(a)

(b)

(c)

(d)

Figure 10.2 Screens from VOYAGE OF THE MIMI. In the "Lost at Sea" example, students use simulated instruments and electronic beacons to find their exact position at sea through a process of triangulation.

of science are studied, including marine science, oceanography, Earth sciences, physics, and navigation. Students watch, read, interact with the computer, and make decisions to solve real problems associated with the television series.

The initial success of the series has been so great that a SECOND VOYAGE OF THE MIMI has been completed. In addition, other programs using the strengths of the different media and integrated into a well-documented guided discovery lesson are being produced.

CD-ROM DISKS

The availability of read-only CD-ROM disks is another tool that has just started to reach the science education market. To date, publishers have used the CD-ROM primarily to store reference material resembling encyclopedias. Typically, a CD-ROM can store the resources of a large set of books. Data can be stored in almost any form—including text, digitized images, voice, and musical notes. A single 12-cm disk holds over 550M of data—about 270,000 pages of typewritten text. Using the extraordinary focus of the laser, the disk can be read with great precision, and the information does not deteriorate. Because graphics and sound capabilities are present, new developers will use a much wider variety of applications.

SCIENCE HELPER K–8 (PC-SIG) (Figure 10.3) contains nearly 1,000 retrievable lesson plans for science units. These lesson plans were developed over a fifteen-year period with support from the National Science Foundation. They were originally a part of the SCIS (Science Curriculum Improvement Study), SAPA (Science—A Process Approach), COPES (Conceptually Oriented Program for Elementary Science), ESSP (Elementary Science Study Project), USMES (Unified Science/Math for Elementary School), MINNEMAST (Minnesota Math/Science), and ESS (Elementary Science Study) programs. Lessons can be viewed on the computer screen or printed on a standard printer. The disk provides a multitude of hands-on/minds-on lessons for every elementary science topic. The system requires an IBM PC or compatible computer and a CD-ROM player.

GEOVISION MAPS (Geovision) lets users recall thousands of maps in various scales and configurations (Bitter, 1988). Teachers and students can easily examine geographic characteristics or population density, plot routes, or calculate distances. Maps can be used for such projects as comparing countries that have an abundance of specified minerals, examining coastline and oceanographic problems, and investigating the possible environmental impact of specified human activities.

Science students frequently consult encyclopedias for various research projects. Encyclopedias have already begun to appear in CD-ROM format. These offer the user the ability to search all volumes simultaneously for any article that mentions the desired topic. Unfortunately, early CD-ROM encyclopedias offered few enhancements beyond the basic printed page. Students often become discouraged with this "print-on-screen" approach and complain that they are sometimes inundated with information they don't want, just because an otherwise irrelevant article contained a relevant keyword. Most students seem to prefer the traditional hardcopy encyclo-

```
************* Query Data **************        *
* # of Lessons chosen: 16              * >>>>> Main Menu  <<<<<
*                                      *
* Grade:                               * A.  pick Grade(s)
*      Grade 5                         * B.  pick Curriculum PROGRAM(s)
* Subject(s):                          * C.  pick Academic SUBJECT(s)
*      Physics                         * D.  pick CONTENT Theme(s)
* Content(s):                          * E.  pick PROCESS(es) of Science
*      Energy                          * F.  pick LESSON TITLE
* Process(s):                          * G.  pick BOOK TITLE
*      Experimenting                   * H.  pick LESSONS to print
*                                      * I.  pick Vocab Words or phrases
*                                      * J. >view Lesson Titles
*                                      * K.  view Program Philosophy
*                                      * L.  EXIT the program
*                                      *
*                                      *
*                                      *
*                                      *
*                                      *
*                                      *
*                                      *
*******************************************
   End letters=move Return=select Esc=cancel &=and/or del=delete ?=help
```

(a)

```
   LESSON TITLE (Abstract): The Compass

   LESSON TITLE (Source): The Compass

   BOOK TITLE: Batteries And Bulbs: Circuits/Magnets

   PROGRAM NAME: ESS

   DESIGNATED GRADE LEVEL: 2-6        RATED GRADE LEVEL: 4-6

   STARTING PAGE: 003                 ENDING PAGE: 009

   SUBJECT: Physical Science, Physics, Chemistry

   PROCESS THEME: Experimenting

   CONTENT THEME: Systems, Energy

   KEYWORDS: electricity, electromagnetic field, magnetism, compass



   PgUp=previous page PgDn Return=next page Esc=cancel F9=view lesson
```

(b)

```
   COMPASS

        This lesson includes information on the operation of a
   compass for the teacher and ways in which the needle of a compass
   can be deflected.  Placing a compass near the wire of a completed
   electrical circuit will produce a deflection in the compass
   needle, as will placing the compass near another compass or near
   a paper clip.  There are several interesting activities for the
   children using compasses, including a set of predictions
   regarding the behavior of the compass needle.




   PgUp=previous page PgDn Return=next page Esc=cancel F9=view lesson
```

(c)

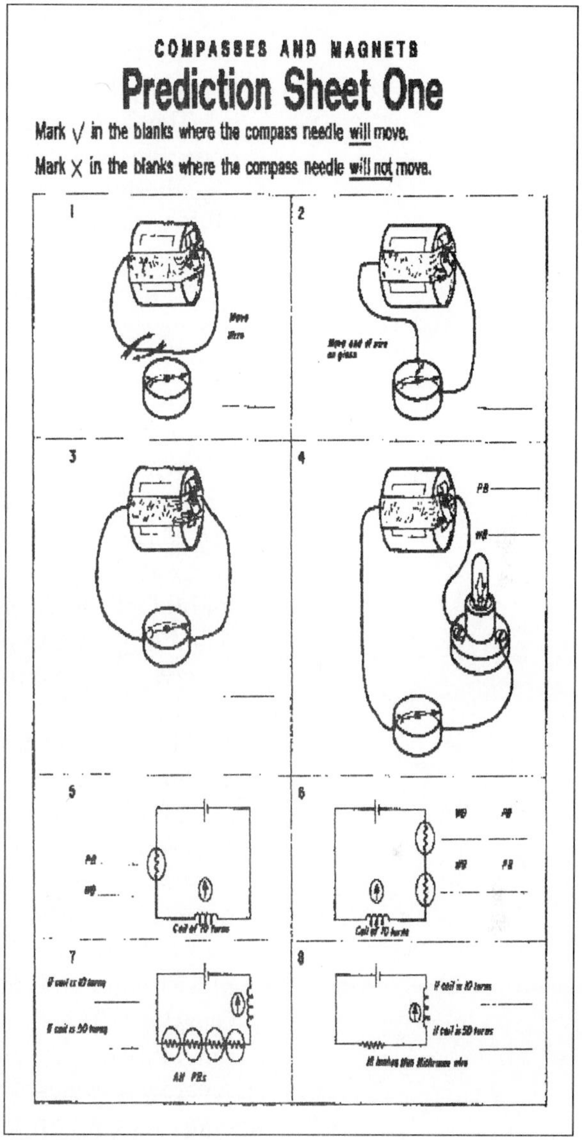

(d)

Figure 10.3 SCIENCE HELPER K–8 helps teachers locate and develop lesson plans for units of instruction in science. Screens (a–c) show part of the search process. Figure (d) is a printed diagram generated by the computer for easy teacher use.

pedia, which makes it easy to browse, to follow a natural path in searching for information, and to examine pictures and tables relevant to the topic they are pursuing.

The Compton subsidiary of *Encyclopaedia Britannica* currently offers an entirely new software product that is likely to revolutionize our methods of gathering information. The COMPTON'S MULTIMEDIA ENCYCLO-PEDIA is a complete encyclopedia on a 585M CD-ROM disk. However, it goes beyond the simplistic approach employed in earlier computerized encyclopedias. Besides its 9 million words of text, the disk includes 15,000 photographs, sixty minutes of audio, a U.S. history timeline, a world atlas, science feature articles, an on-line dictionary, and a built-in note pad so that students can take notes from articles and print them for later use. In addition, the program permits the user to access information in several different ways. For example, learners can use an improved keyword strategy comparable to that available in the earlier computerized ency-clopedias to find all articles in the entire encyclopedia that mention a designated topic or combination of topics; but they can also follow more traditional approaches, such as moving quickly from one topic to another without having to pull another volume off the shelf or moving from a general topic to a more specific topic as the need and interest naturally dictate. The program provides seven different ways to access the information on the CD-ROM disk and permits the user to move easily from one access method to another.

CD-ROM technology has barely begun to have its impact on science education. Within the decade, electronic encyclopedias will be found in most school libraries, and students will use these to access information much more efficiently than has traditionally been possible. However, simple logistics will make it unlikely that electronic encyclopedias will replace hardcopy versions. If a school has just a single copy of an electronic encyclopedia, only one student (or small group of students) can use it at any time; this may be much less efficient than having several students use separate volumes of a hardcopy encyclopedia. The encyclopedia can be put onto a local area network (LAN), enabling several computers to access the same CD-ROM disk; but when this is done, the entire LAN must be dedicated to the single purpose of using that one CD-ROM disk. In actual practice, schools are still likely to retain the traditional hardcopy version of the encyclopedia, so that students will start a search at the computer and then continue at a more leisurely pace at the hardcopy encyclopedia. In fact, as CD-ROM encyclopedias become more popular, it is likely that students will become more enthusiastic about the use of encyclopedias in general, and schools may need two or three hardcopy versions of the encyclopedia to satisfy the insatiable needs of their eager learners.

In addition to the COMPTON'S MULTIMEDIA ENCYCLOPEDIA, there are other CD-ROM encyclopedias. THE NEW GROLIER ELECTRONIC ENCYCLOPEDIA includes more than 30,000 articles and thousands of pictures, diagrams, and graphs. It lacks the sound, animation, and easy access strategies of COMPTON'S. World Book's INFORMATION FINDER offers most of the text of the *World Book Encyclopedia* and much of the *World Book Dictionary* on a CD-ROM disk. However, it does not offer graphics, sound, or the easy access strategies of COMPTON'S. Compared to COMPTON'S MULTIMEDIA ENCYCLOPEDIA, Grolier's major advantage is lower price; World Book's major advantage appears to be the popularity of the hardcopy *World Book Encyclopedia*. It can safely be assumed that these and other vendors will imitate and expand upon the successful strategies employed in COMPTON'S. Science teachers can expect continued, rapid, attractive developments in computerized delivery of traditional library resources like encyclopedias.

Besides encyclopedias, other reference materials are likely to appear on CD-ROM with interfaces that make them useful for science students. For example, such resources as *The Handbook of Chemistry and Physics*, *Standard Methods of Chemical Analysis*, *Wild Flowers of the United States*, *Gray's Anatomy*, and the *Handbook of Microbiology* contain vast quantities of information that would be of interest to science students but at present are difficult to access. A *Science and Technical Reference Set* on CD-ROM was recently published by McGraw-Hill. It contains the *Encyclopedia of Science and Technology* and the *Dictionary of Scientific and Technical Terms*, including over 7,300 articles, 98,500 terms, and 115,500 definitions. For these reference materials to become truly accessible, they must not only be entered onto CD-ROM disks, but they also must be integrated with useful access systems (perhaps employing a hypermedia format) to let students find information as their needs and interests dictate.

ARTIFICIAL INTELLIGENCE

Tutorials providing branching programmed discussion as a form of CAI are thoroughly discussed in Vockell and Schwartz (1988, Chap. 1) and briefly in Chapter 2 of this book. It is possible to combine branching tutorials with artificial intelligence to provide a strategy called intelligent computer-assisted instruction (ICAI). Like a good human tutor, an ICAI program monitors student responses and tries to determine reasons behind student errors. The computer then provides tutorial or remedial information specifically designed to overcome these errors.

A good example of this strategy is a program called MENDEL, which tutors high school and college students on various aspects of genetics. MENDEL is both a problem-solving simulation and an intelligent tutorial. In the first of MENDEL's roles, the computer presents the learner with problems to solve, while using the computer as a tool to conduct experiments and collect or interpret data. In its second role, MENDEL assumes the role of a human tutor. This program is described in detail in Vockell and Schwartz (1988) and in Streibel and co-workers (1987). Applied to reading, such a program could perform the following tasks for a student studying a reading passage:

- Make inferences about the answers given by the student, drawing conclusions about what skills the student possesses or still needs to develop.
- Maintain a history of a student's actions (including the strategies the student tried and the results of these strategies).
- Make inferences about the reasons for the student's problem-solving actions. These are drawn from a combination of what the student has done and said. (In doing this, a human tutor is building a model or representation of each student's or group of students' comprehension abilities.)
- Compare a model of a student's knowledge with the tutor's understanding of the problem.
- Make a decision on the form of tutorial advice and the timing of this advice.
- Evaluate whether the student has benefited from the advice.

When students run such a program, their experience is much the same as if they were reading and thinking about a passage with a human tutor who has acute knowledge of comprehension skills and willingness to help whenever help is needed. The development of this program is described in greater detail in Streibel and colleagues (1987). At the time this book went to press, a program like this does not actually exist, but the technology is already available. Similar programs can be developed in other areas where the computer can be programmed to solve a problem and then check to see how closely its solution compares to that of a human learner.

TELECOM-MUNICATIONS: "THE ELECTRONIC SCHOOL"

Not far from where this book is being written is an excellent example of the application of computer and telecommunication technology to instruction. As a result of a five-year strategic plan, an integrated voice/video/data system was developed by the Penn-Harris-Madison School Corporation of Mishawaka, Indiana, in partnership with AT&T to deliver instruction from both near and distant sources, using state-of-the-art technology.

Each classroom in the high school has a telephone, an LAN connection, and a video connection. A modem lets teachers and students access CNN's "Newsroom" and "Week in Review," as well as the National Geographic KIDNET, the Indiana Department of Education Access Network (IDEAnet), and a host of other telecommunications resources. The video connection lets the classroom teacher schedule and control the delivery of media from almost any source including satellite, cable TV, and in-house video circuit. To use the media, the teacher merely pushes the button for the desired function, and source machines give the teacher complete control of the devices as if they were in the classroom.

The data connection in each room allows interfacing with a number of LANs for various instructional and management resources. For example, an instructional management system with student permanent record information, an electronic gradebook, and a pupil mastery learning profile can be accessed by both teachers and administrators through an LAN. A variety of computerized instructional software is also available via the distributed data network. In addition, Penn High School science teachers can regularly conference with other science teachers across the state through the IDEAnet bulletin board.

Cables, switching, and control devices are expandable, and all control functions are software-driven so that changes in the program can be easily accomplished.

Besides the integrated system, each school has desktop video-editing equipment, so that video shots by students can be combined with the "Video Encyclopedia of the 20th Century," recorded video from both CNN "Newsroom" and "Week in Review," and computer-generated images such as animations, scrolls, texts, graphic designs, and original music to produce student video reports. Over 360 student video projects were completed during the 1988–1989 school year. For example, an Earth science class did a thirty-second ecology commercial. Biology and chemistry classes include video production in their units of study. Through the use of laserdiscs and HYPERCARD, staff and students are turning desktop programs into interactive video programs.

Finally, a state-of-the-art planetarium includes a Star Machine that allows the planetarium director to show the night sky for any date. The Star Machine is augmented by more than 200 special-effects projectors. Technologies such as CD-ROM, CD-audio and video, and interactive laserdisc display video images on the planetarium dome and generate music and sound. This gives the director the ability to develop his own videos for use in the planetarium.

Not every school needs the exact set of technological equipment employed at Penn-Harris-Madison. However, with resources like these, the interaction among students, teachers, administrators, and outsiders can create "schools without walls" with an incredibly rich environment. A potential exists for schools and science classrooms in our country to be connected to each other and to remote places and resources so that distance learning can be commonplace by the end of the century. The classroom can be in Kansas and the resource teacher and laboratory can be in Alaska. Combining satellite transmissions with computer networks makes it possible to go anywhere to learn anything, an exciting prospect. If it is to become reality, the type of planning and cooperation in the "Schools of the Future" project must become commonplace in the boardrooms and schools of our nation.

SUMMARY

We have taken the position that computers don't teach complex skills; teachers teach these skills, using the computer as a tool. As artificial intelligence and intelligent computer-assisted instruction evolve, the roles of the teacher and the computer may shift. It is possible that within the next ten years *some* science skills may be most effectively taught by having students interact directly with computers without teacher mediation. However, the overall guiding principle for teaching science skills in the classroom will continue to be that a knowledgeable teacher should stimulate a motivated group of students to pursue these skills, using computers or other materials as appropriate tools.

The future is likely to see expanded applications of the computer in the field of science instruction. In many cases, the technology is already here, and we merely await more widespread adaptation and implementation of existing technology. In addition, technological advances will make possible some innovations in science instruction. Hypertext will make it possible to provide information (such as definitions of terms) at the time and in the order it is really needed. Interactive video will make it possible to integrate textual and visual presentations and for learners to move easily backward and forward within a multimedia package. Intelligent computer-assisted instruction will make it possible for the computer to respond more precisely to the specific needs of individual learners as they are experiencing computerized reading instruction.

We also expect to see more widespread implementation of well-integrated, multimedia approaches to science education. Publishers are also likely to expand their efforts in producing good materials for implementing the hands-on/minds-on approach to science education. Because computer technology is still in a stage of rapid development, it is likely that other innovations will lead to instructional developments that cannot be predicted at the present time.

Finally, it seems inevitable that combining satellite transmissions with computer networks will make it possible to go anywhere to learn anything. "Schools without walls" may become commonplace with the assistance of computers and telecommunication technology. The learning environments in such schools have the potential to be incredibly rich, allowing for a diverse set of natural phenomena, people, technology, and cultures to be explored by students at all levels. This potential will not be realized without careful planning and thoughtful implementation.

REFERENCES

Bitter, G. "CD-ROM Technology and the Classroom of the Future." *Computers in the Schools*, 5(1/2) (1988): 23–34.

Daynes, R. "Who, What, When, Where, Why and How Much of Videodisc Technology." In R. Daynes and B. Butler (Eds.), *The Videodisc Book: A Guide and Directory*. New York: Wiley, 1984.

Sherwood, R. D. "Optical Technologies: Current Status and Possible Directions for Science Instruction." In J. D. Ellis (Ed.), *Information Technology and Science Education*, 1988 AETS Yearbook. Columbus, Ohio: ERIC, 1989.

Streibel, M. J., et al. "MENDEL: An Intelligent Tutoring System for Genetics Problem-Solving, Conjecture, and Understanding." Paper presented at the annual meeting of the American Educational Research Association, Philadelphia, May 1987.

Vockell, E. L., and E. Schwartz. *The Computer in the Classroom*. Watsonville, Calif.: Mitchell, 1988.

CHAPTER *11*

SOFTWARE REVIEWS

THIS CHAPTER PROVIDES FAIRLY detailed reviews of fifty software packages available for the science curriculum. Appendix B provides less detailed summaries of a much larger number of programs. The goal here is to discuss those programs that are typical of the kinds of help teachers can expect from the computer in their science classrooms. These programs cut across grade levels and subject matter. The programs described in this chapter may also be discussed elsewhere in this book, and cross references can easily be found in the book's index.

TITLE: ADVENTURES WITH CHARTS AND GRAPHS: PROJECT ZOO

Company: National Geographic Society
Grade Level: 3–5
Computer: Apple II
Required Memory: 64K
Special Equipment: None

SUBJECT MATTER
Graphs and tables
Characteristics of animals
Animal classification
Characteristics of zoo animals
How zoos meet animals' needs
Survival needs of animals
Principles of zoo design

OBJECTIVES
- Use picture graphs, bar graphs, and tables to record and interpret information
- Solve word problems
- Apply creative problem solving to the design of a zoo

PROGRAM DESCRIPTION
PROJECT ZOO is a multimedia courseware adventure in mathematics and science. Built around a zoo theme, it develops skill in graphing, measurement, mapmaking, and research; promotes understanding of the characteristics and survival needs of animals; and provides practice in problem solving. Students begin as visitors to a zoo and end by designing zoos of their own. Three programs form the heart of the package. "Zoo Goer" is an interactive lesson in which students visit animal houses in a zoo, collect information, and learn how to represent the data they have collected in picture graphs, bar graphs, and tables. "Zoo Collector" is a game in which students determine the identities of "mystery animals" by interpreting data from graphs and tables. "Zoo Builder" guides teams of students in designing their own zoos. The program divides the complex task into a series of steps, guides students, records information, allows students to

manipulate their designs on the computer screen, and offers feedback on how well their decisions meet the needs of animals, visitors, and staff. Other components of the kit include a filmstrip with background information on zoos and the needs of animals and a student book, *The Zoo Animals Fact Book*, which provides specific information on animals students select for their own zoos.

EVALUATION

The format of this multimedia program is highly motivating for teaching students concepts related to zoo animals, scientific problem-solving skills, and mathematical graphing skills. The program moves from the concrete to the abstract, from building skills to enabling students to use these skills to solve problems of increasing complexity. The teacher's guide is helpful and provides suggestions for using the wealth of material in constructive ways. The computer programs and accompanying materials are well integrated, colorful, and appealing. This program could be used by elementary teachers in preparation for a zoo visit or as a follow-up to a visit. In either case, it could help make the visit much more purposeful and productive.

TITLE: AIR TRACK SIMULATOR

Company: CDL
Grade Level: 10–adult
Computer: Apple
Required Memory: 48K
Special Equipment: None

SUBJECT MATTER
Physics
Conservation of energy
Conservation of momentum
Elastic and inelastic collisions

OBJECTIVES
- Observe and analyze collisions on a horizontal, frictionless surface
- Develop a conceptual understanding of elastic and inelastic collisions
- Derive mathematical relationships among mass and velocity of two objects that collide
- Predict velocity and direction of movement in elastic and inelastic collisions

PROGRAM DESCRIPTION
The theory and tutorial sections of the program cover the concepts of conservation of energy, conservation of momentum, and collisions. The program derives equations and defines concepts and symbols. Students can observe animated graphic representations of two masses colliding on an air track. The masses, direction of movement, and changes are recorded, and the movements can be observed. Following observations, students can predict events from various setups; and the tutorial will correct and explain, if students are incorrect in their calculations. Students can also design unique collisions by varying masses, velocities, directions, and types of collisions.

EVALUATION
This program can be used both as a lecture demonstration and for individual study. The theoretical explanations are short and well explained.

The tutorial section allows students to correct answers once and will explain the calculations if the students cannot enter the correct answer. The simulations are good, and "cars" can be stopped in order to write down information at various times during the experiment.

TITLE: ALL ABOUT MATTER

Company: Ventura Educational Systems
Grade Level: 4–6
Computer: Apple
Required Memory: 48K
Special Equipment: None

SUBJECT MATTER
Chemistry
Atoms and molecules
States of matter

OBJECTIVES
- Learn basic information about atoms, molecules, and states of matter
- Visualize a simple model for atoms and molecules
- Practice vocabulary related to these concepts

PROGRAM DESCRIPTION

Students can select one of three topics. Each topic has a "lesson," a "probe," "games," and "quizzes." A glossary of terms can be accessed at any time. Each lesson is a short, self-paced presentation of information in text windows about a graphic appearing on the screen. The probe provides the vocabulary, with the term identified on the graphic and then associated with other key words. Games include word scrambles and searches.

EVALUATION

The program presents a great deal of information without much activity to maintain interest. The pieces of the program are short enough, and the visuals help keep students on task. The probe is a good way to review vocabulary, presenting a visual clue as a mnemonic to help students remember information. The program could be used to introduce concepts but would be better in a review mode. It is also better for one student than for a group. The program will help develop a working vocabulary and provide a simple, visual model; the presentation, however, is very similar to a textbook presentation and will not help with the very abstract nature of matter.

TITLE: ANIMAL LIFE DATABASES (ALSO ASTRONOMY; SPACE; CLIMATE AND WEATHER; ANIMAL LIFE; ENDANGERED SPECIES)

Company: Sunburst
Grade Level: 4–12
Computer: Apple II
Required Memory: 64K
Special Equipment: None

SUBJECT MATTER

Physical characteristics, classification, behavior, food, habitat, reproduction, adaptations, and so on of the species or topic under consideration

Common and scientific names, physical characteristics, trophic level, diet, habitat, and so on of the species or topic under consideration

Fiction and nonfiction books about the topic under consideration

OBJECTIVES

- Develop skill in using databases, including deriving information from an existing database, entering information into a database, and designing databases from scratch
- Develop skills related to scientific record keeping
- Use simple "and/or" logical operators to search and sort information in a database
- Use database information to look for patterns, make inferences, solve problems, or test hypotheses
- Enter information into a database in a systematic manner suitable for retrieval and scientific use

PROGRAM DESCRIPTION

Learners use BANK STREET SCHOOL FILER to manipulate databases on the topic under consideration (e.g., animal life or endangered species). Guidelines lead students through stages of first using existing databases to answer questions, then entering information into existing databases, and finally developing databases of their own. Students are encouraged to form inferences and test generalizations.

EVALUATION

This database management system is not as powerful as those with which some teachers (and many students) will be familiar. However, learning and employing simple English commands for searching and sorting is easy. The accompanying manual provides guidelines to encourage students to apply database skills in a variety of settings. The hierarchical approach (first browsing, then entering data, and finally developing databases) can be easily and effectively adapted to a learning cycle approach, each step corresponding to a phase in the learning cycle.

TITLE: ASTRONOMY: STARS FOR ALL SEASONS

Company: Educational Activities, Inc.
Grade Level: 4–12
Computer: Apple, TRS-80, Atari, PET
Required Memory: 64K
Special Equipment: None

SUBJECT MATTER
Astronomy
Rotation and revolution
Ascension and declination

OBJECTIVES
- Visualize the night sky as seen from a backyard
- Learn the concepts of rotation and revolution
- Learn to locate objects in the sky using right ascension and declination

PROGRAM DESCRIPTION
The program introduces the concepts of rotation and revolution in relation to the changing sky. The tutorial explains the appearance of the sky and why the stars and constellations we see change from day to day. Graphical views show the nighttime sky as seen from a backyard.

EVALUATION
The screen explanations are in large print. The explanations of rotation and revolution and the changing nighttime sky are simple and more of a "this is how it is" than a scientific explanation. The relationships shown in the visuals are somewhat distorted, and the use of right ascension and declination is beyond the grade level for which this program is intended.

TITLE: BAFFLES

Company: Conduit
Grade Level: 6–adult
Computer: Apple, IBM
Required Memory: 48K
Special Equipment: None

SUBJECT MATTER
Problem solving
Logic and inference
Black box experiments

OBJECTIVES
- Understand the process of indirect observation and develop the ability to infer a physical pattern from effects that cannot be seen
- Use a series of indirect observations to formulate and test hypotheses
- Keep accurate records from scientific investigations and use these to develop hypotheses and draw conclusions

PROGRAM DESCRIPTION
This is a simulation of a "black box" experiment. The learner has a "box" that has deflectors or baffles inside, which the learner cannot see. Probes are shot into the box, and students must observe both the entry and exit points to infer the location and slant of the deflectors.

EVALUATION
The format is challenging, and the supplementary materials suggest strategies for integrating the program into the curriculum. Black box programs are useful for teaching students that scientists cannot always "see," but they can still develop models based on their experiments and inferences. This program is a good example of content-free software that focuses on higher-order thinking skills and that can be incorporated into the science classroom. A very large number of similar programs, such as THE INCREDIBLE LABORATORY and SAFARI SEARCH are summarized in detail in Vockell and van Deusen (1989).

TITLE: BALANCE

Company: Diversified Educational Enterprises
Grade Level: 7–12
Computer: Apple II, IBM
Required Memory: 48K
Special Equipment: None

SUBJECT MATTER
Predator/prey relationships
Environmental variables that affect predator/prey populations
Carrying capacity

OBJECTIVES
- Develop skill in interpreting and evaluating data
- Develop skill in manipulating variables
- Develop insight into decision-making processes
- Given data related to a predator/prey problem, make appropriate inferences, develop plausible hypotheses, and develop an appropriate experiment to test those hypotheses

PROGRAM DESCRIPTION

This is a student-interactive program designed to investigate predator/prey relationships. Using wolf and deer populations, students vary population sizes, environments, and deaths. Manipulation of these variables permits study of the interrelated populations. The influences of carrying capacity, food supply, environmental conditions, and external pressures on the two populations are investigated. The teacher's guide includes background on the model used to simulate the predator/prey relationship, suggested teaching strategies, a student laboratory guide, and a problem-solving test keyed to the program's objectives.

EVALUATION

Tabular and graphic output clearly illustrate effects of variables on the related population. Students can redo tables and graphs quickly. This is the kind of experiment that is difficult to do in the biology laboratory. Student interest is high with the approach used. The documentation provides the teacher with a great deal of support in integrating the package into the curriculum. The program can be easily adapted to all phases of the learning cycle.

TITLE: BIOLOGY: REPRODUCTIVE SYSTEM, RESPIRATORY SYSTEM, CIRCULATORY SYSTEM (AND OTHERS)

Company: Prentice-Hall Courseware
Grade Level: 9–12
Computer: Apple II
Required Memory: 64K
Special Equipment: None

SUBJECT MATTER
Reproduction in animals and humans
Respiratory system structure and function in animals and humans
Circulatory system structure and function in animals and humans

OBJECTIVES
- Identify and describe the structure and functions of the respiratory and circulatory systems in various animals and humans
- Identify and describe how various lower animals accomplish reproduction
- Identify the structure and function of the male and female reproductive systems in humans

PROGRAM DESCRIPTION
Each tutorial program in the Prentice-Hall Biology Courseware Series has the same structure. In the "Preview" section, all standardized screen directions and symbols are shown and explained. Students can use this section to see the on-line glossary, the program's section and sequence titles, and the objectives for each. The "Instruction" section presents graphic-augmented instruction. Students can control the sequencing and pacing of instruction. The "Evaluation" section includes the "Test," "Reteach," and "Record-Keeping Systems." Students can choose to take an end-of-section test, or they can take a combined test after they complete the entire program. Following the test, students receive an automatic "Reteach" on each test item that they answered incorrectly. While in the reteach mode, students can access the help menu or the glossary. The "Record-Keeping System" holds up to 100 student scores and has print capability.

EVALUATION

This series of programs in biology is a good example of a flexible, menu-driven, well-organized, conceptually based tutorial on topics of importance in animal biology. Although the instruction is linear, it is illustrated with graphics and animation. Sound provides cues to students, and questions test understanding. Help screens are also available on student demand. Once they master the commands for moving around in the programs, students can progress rapidly through all segments and sections because the commands are standardized. The objectives for the various segments are well written, and the instruction and tests are in alignment with them. The teacher's guide is helpful with background information, suggested discussion questions, a test, a glossary, blackline activity masters, and a record-keeping chart. An irritating problem for students is the necessity for disk switching with $5\frac{1}{4}$-inch disks on a one-drive or even two-drive system.

TITLE: BIRDBREED

Company: EduTech
Grade Level: 9–12
Computer: Apple II
Required Memory: 48K
Special Equipment: None

SUBJECT MATTER

Inheritance of genetic traits
Genotypes and phenotypes
Analysis of genetic crosses
Models of inheritance
Genetic probability

OBJECTIVES

- Make inferences for underlying genotypes from phenotypic information and breeding results
- Develop skill in the analysis of results of crosses
- Develop skill in the formulation of genetic models to explain observed outcomes
- Develop skill in the testing of models and hypotheses
- Develop understanding of the concept of probability as it relates to small numbers of offspring

PROGRAM DESCRIPTION

This program provides students with a group of breeding parakeets and lets them carry out breeding investigations with user-selected birds from among sixteen groups of birds with defined phenotypes. The phenotypic appearance of the resulting offspring forms the output for the program. Students can accumulate up to 100 birds for the analysis of genotypes and patterns of inheritance. Clutch sizes vary from four to eight fledglings, incorporating the variations observed in small populations in nature.

EVALUATION

This program provides a detailed and comprehensive simulation of inheritance patterns involving one or two traits. The principle of probabilities in genetics, especially with small clutch sizes, is clearly demonstrated. Four levels of difficulty make students familiar with single-gene dominance, incomplete dominance, sex linkage, multiple alleles, and linkage with cis or trans arrangements. The guide suggests that the simulation should be used in conjunction with real laboratory investigation—a valuable suggestion. This program could be effectively used in all phases of the learning cycle.

TITLE: BOTANICAL GARDENS

Company: Sunburst
Grade Level: 6–12
Computer: Apple II
Required Memory: 48K
Special Equipment: A mouse is suggested

SUBJECT MATTER
Plant growth
Environmental factors affecting plant growth
Experimental design

OBJECTIVES
- Collect and average experimental data for one plant for ten different control settings
- Graph a series of points on a two-dimensional graph
- Write a descriptive statement that depicts the plant's response to the changed setting
- Experiment, controlling one variable at a time
- Interpret data presented in order to infer the best conditions for the growth of the selected plant

PROGRAM DESCRIPTION
As students "grow plants" in a computerized environment, they must identify and manipulate appropriate variables to come to a proper understanding of the growth requirements for a selected plant. The program also includes a "genetics" laboratory," in which students can design their own custom seeds. The plants from these seeds can be saved on disk to challenge other students.

EVALUATION
This format is highly motivating for teaching students concepts related to plant growth and integrated scientific problem-solving skills. Students develop such strategies as conservative focusing, analyzing, making organized lists, and looking for patterns or sequences. This program can easily be adapted for use in a learning cycle, especially in group activities during the exploration and concept application phases.

TITLE: CHEM LAB

Company: Simon & Schuster
Grade Level: 4–9
Computer: Apple II, IBM
Required Memory: 64K
Special Equipment: None

SUBJECT MATTER
Properties of elements and compounds
Periodic chart
Chemical formulas
Chemical reactions
States of matter

OBJECTIVES
• Design and carry out chemical reactions to produce a certain target compound, by using given raw materials and selecting the conditions under which the experiment is to be conducted

PROGRAM DESCRIPTION
This program puts a chemistry set inside the computer. It allows students to conduct fifty different experiments in a simulated robot laboratory with thousands of possible combinations of chemicals. The reactions in the program are based on authentic chemistry principles and mimic actual reactions, many of which are too costly or too dangerous to do in the real classroom laboratory.

EVALUATION
This interesting program allows students to capture the flavor of scientific exploration in a safe environment. It would be an excellent alternative to a real chemistry set in an unsupervised setting at home for elementary and middle school students. It could be useful in the classroom to introduce students to *qualitative* aspects of chemical reactions, the properties of elements, and common chemical reactions, especially during the exploration phase of the learning cycle. It would not be useful during concept introduction because of the time it would take for the teacher or students to use the robot laboratory to demonstrate a particular reaction. It could also be useful to introduce students to relationships found in the periodic table and to chemical formulas.

TITLE: CHEM LAB SIMULATION 1: TITRATIONS (ALSO CHEM LAB SIMULATION 2: IDEAL GAS LAW; CHEM LAB SIMULATION 3: CALORIMETRY; CHEM LAB SIMULATION 4: THERMODYNAMICS)

Company: High Technology Software Products
Grade Level: 10–adult
Computer: Apple, Atari
Required Memory: 48K (Apple); 32K (Atari)
Special Equipment: Color monitor helpful

SUBJECT MATTER
Chemistry
Acid/base titrations
Determination of equilibrium constant
Determination of Avogadro's number
Simulations 2: Ideal gas law and kinetic molecular theory
Simulations 3: Hess' Law, heats of reaction
Simulations 4: Heat of vaporization and thermodynamics of equilibrium reactions

OBJECTIVES
- Perform simulated experiments and gather data
- Perform a simulated acid/base titration
- Determine the equilibrium constant of a weak acid
- Determine Avogadro's number using a simulated titration and mono-molecular experiment

PROGRAM DESCRIPTION
Students titrate on-screen, with the indicator changing color as the endpoint approaches. They control how fast the solution enters the beaker, so that the endpoint can be reached exactly. The program series also includes IDEAL GAS LAW, CALORIMETRY, and THERMODYNAMICS. All have high-resolution graphics to help visualize the chemical concepts being explored.

EVALUATION

The programs could be used for laboratory preparation or as practice. The manuals explain the procedures used on the disks as well as some of the theory behind the simulations. These programs are useful in situations where the laboratory experiment can actually be performed; but they add a tremendous amount to a class that cannot actually complete the laboratory with real equipment. Some of the simulated instruments are difficult to read, but users can practice and get results with reasonable accuracy.

TITLE: CHEMAID

Company: Ventura
Grade Level: 6–12
Computer: Apple
Required Memory: 48K
Special Equipment: None

SUBJECT MATTER
Chemistry
The elements
Retrieval of information from a database

OBJECTIVES
- Learn names and characteristics of the elements
- Identify elements and learn to spell their names
- Recognize family position on the periodic table
- Identify patterns within families, groups, or periods

PROGRAM DESCRIPTION
This program has both an identification game and a data-retrieval utility. The elements are organized on a colorful periodic chart. Certain elements are erased, and students must identify each symbol, name, and atomic number and spell each name correctly. In the data-retrieval option, students can scan by element, ask for a specific element from a variety of characteristics, or look at the periodic table. Electron configuration, specific gravity, and other special properties are in the database.

EVALUATION
This is a good electronic database to search for various properties of elements. It is possible to locate all elements that have a specific property or to identify one element from some of its characteristics. A teacher could use this program in an inquiry mode, having students identify particular mystery elements; or real elements could be observed, and the students could use the database to identify the name of each element. The program also provides drill and practice and would be useful in establishing a familiarity with the elements.

TITLE: CHEMISTRY ACCORDING TO ROF

Company: Richard O. Fee
Grade Level: 10–adult
Computer: Apple
Required Memory: 48K
Special Equipment: Paddles useful for some programs

SUBJECT MATTER
General chemistry

OBJECTIVES
- Introduce concepts in general chemistry
- Apply the concepts learned in general chemistry
- Review the mathematics necessary to solve chemistry problems
- Demonstrate principles of chemistry

PROGRAM DESCRIPTION
This series consists of seventeen disks designed to provide drills, simulations, review, and tutorial help in a general chemistry course. For example, one program simulates the "Millikan Oil Drop" experiment, and another the ideal gas laws. They do not cover nuclear or organic chemistry. Help is provided, sometimes with graphics and animation. The program keeps track of student progress.

EVALUATION
These programs can be used in many different ways in the classroom. Some programs are useful for whole-class demonstrations, while others are useful to help students learn to solve specific types of problems. Some of the simulations are appropriate for pre- and postlaboratory discussions. The programs provide step-by-step explanations, and these are useful as a teaching tool or for review. The programs are not flashy but contain good questions that require matching rules, doing problems in different ways, and expressing answers in proper format. The disks can also be used by individual students for preparation, review, or making up work.

TITLE: CHEMISTRY: THE PERIODIC TABLE

Company: MECC
Grade Level: 9–12
Computer: Apple
Required Memory: 128K
Special Equipment: None

SUBJECT MATTER
Chemistry
The elements
Retrieval of information from a database

OBJECTIVES
- Gain knowledge about the elements
- Understand the periodic nature of the table of elements
- Search databases for specific information

PROGRAM DESCRIPTION
The program includes a "Property Search" and an "Element Inquiry." In the "Property Search," students can look up data about the ninety-two naturally occurring elements. They can search by element, group, or period. This part of the program is designed to acquaint students with properties of the elements. "Element Inquiry" has two levels of difficulty. Students must collect property data to match an unknown element. An element is missing from the periodic table, and students must determine its atomic number, oxidation state, electron configuration, atomic mass, and state. This is done by searching the databases and the colorful periodic chart. Neighbors that have related properties are located to help infer the correct properties.

EVALUATION
The database retrieval contains almost all information students would wish to know about the elements. It is easy to use, and patterns can be inferred by collecting data. The method is inquiry-oriented and challenges students to apply knowledge of the periodic table. The "Element Inquiry" is a good application of knowledge gained. For example, students must know how to determine what the atomic number is from looking at the position on the periodic chart. This program would be useful in identification of real elements from observed properties.

TITLE: THE DESERT

Company: Collamore, D. C. Heath
Grade Level: 4–6
Computer: Apple II
Required Memory: 128K
Special Equipment: Color printer and mouse are suggested

SUBJECT MATTER
Desert plants and animals
Nocturnal and diurnal desert animals
Desert food chains

OBJECTIVES
- Identify desert plants and animals by habitat and name
- Classify desert animals as either nocturnal or diurnal
- Construct desert food chains and food webs
- Create accurate scenes of Sonoran Desert life

PROGRAM DESCRIPTION

This program combines science activities on the computer disk with a science reader. The illustrations in the reader can be generated by the computer program. "Identifying Animals and Plants" contains four scenes from the Sonoran Desert. Students are asked to label the animals in each scene, using the science reader for help. The "Exploring Habitats" activity asks students to place the appropriate animals into a day, night, and sunrise scene. The activity lets students create "problem" scenes with incorrect animals. Another student could then try to determine which animal does not belong in the scene. "Gathering Information" allows students to build small data files with information about the animals and plants of the Sonoran Desert. "Analyzing Food Chains" presents students with the opportunity to identify food chains in the scenes, using their science reader and database to assist them. It also identifies the food webs on the desert and allows students to modify them or create new ones.

EVALUATION

This is a unique program that combines computer simulation of a desert scene with reading and language arts. Students use both the computer program and the accompanying science reader to obtain the information they need to construct accurate scenes of desert life and to infer the interrelationships among the various forms of life. Suggestions are made in the teacher's guide for follow-up science activities. A creative teacher could also use the program as a beginning point for a creative writing, language arts, or art activity, by having students add text to the color pictures generated by the computer, by creating group presentations concerning some element of desert life, or by creating a piece of desert art designed from the computer scenes. A color printer makes the artwork much more pleasing. The mouse makes it easy to place both creatures and labels on the scenes.

TITLE: DICHOTOMOUS KEY TO POND MICROLIFE

Company: EduTech
Grade Level: 3–9
Computer: Apple II
Required Memory: 64K
Special Equipment: None

SUBJECT MATTER
Microorganisms in a freshwater aquatic environment
Protists and cyanobacteria
Habitats of aquatic microlife
Use of microlife as pollution indices

OBJECTIVES
- Use the computerized dichotomous key to identify common microlife in freshwater ponds and puddles
- Use the database to determine which microlife forms are to be expected in a given microhabitat (or a combination of microhabitats) or which exhibit a particular behavior or condition
- Use identified microlife forms to compile data concerning microhabitat population surveys and population dynamics as well as pollution indices for a given field locale

PROGRAM DESCRIPTION
This package is designed to allow users to identify seventy-nine of the most common protists and cyanobacteria in ponds, puddles, infusions, or commercially cultured material from biological supply houses. The key is the backbone of the package. Students enter either 1 or 2 to select the option in the lower portion of the screen that best describes the microlife under investigation. The upper portion of the screen maintains a continuous update of previous key characteristics. Stylized graphics reinforce certain key characteristics. The number 3 can be used to backtrack through the key if a wrong choice is made. Pressing ? brings students to a glossary without disturbing their progress through the key. After students answer a series of questions correctly, a detailed illustration of the organism will appear on the screen. The key words summarizing previous key characteristics appear at the top of the screen. A brief description listing the name, size, and habitat

of the microlife is found on the right portion of the screen. The "Microhabitat Query" section allows students to review all microlife forms from an alphabetical list or to select only those forms that are found in a particular habitat. When selecting forms by habitat, students can list up to six conditions that may exist in that habitat. To be listed, the microlife form must be found in all the conditions listed.

EVALUATION

This program is an example of a well designed, flexible, computerized key that provides students with both framework and resources to key out common microlife quickly and easily. The ability to move both directions in the key and to access the data file and glossary during the process are assets. The ability to review all organisms that live in a particular environment is valuable, especially in conjunction with live laboratory study. After searching the database by using habitat conditions, students would be in a much better position to identify common organisms in their sample of pond water. The alphabetical listing is not particularly useful, except for browsing to get an idea of the construction of the database. It would be good to see other keys for the science classroom constructed on this model.

TITLE: DISCOVER: A SCIENCE EXPERIMENT

Company: Sunburst
Grade Level: 6–12
Computer: Apple
Required Memory: 48K
Special Equipment: None

SUBJECT MATTER
Scientific observation
Scientific method
Record keeping

OBJECTIVES
- Develop skills of scientific observation
- State and test scientific hypotheses
- Identify data that support or refute a given hypothesis
- Identify and control variables

PROGRAM DESCRIPTION

Students are given several hypothetical life forms and must discover how to keep them alive. Within the simulated laboratory, students can observe these life forms, provide them with different foods at different times, and set up barriers and pathways to promote or inhibit interaction among the creatures.

EVALUATION

This simulation is highly motivating for teaching students how to conduct a scientific experiment. It is especially suited for groups in a cooperative learning environment and can be used with whole-class instruction.

TITLE: ENERGY SEARCH

Company: McGraw-Hill
Grade Level: 5–12
Computer: Apple
Required Memory: 64K
Special Equipment: None

SUBJECT MATTER
Earth science
Energy sources: advantages and disadvantages
Risk analysis, decision making, and making predictions based on data
Graphing and interpreting graphs

OBJECTIVES
- Maintain the smooth and successful operation of an energy factory
- Identify the advantages and disadvantages of different energy sources
- Make decisions based on changing content information provided by the computer

PROGRAM DESCRIPTION
Students or groups run an energy factory to provide energy for homes and businesses. They begin with animal power and gradually evolve to water, wood, coal, oil, nuclear, and solar power. Students must learn the advantages and disadvantages of each and improve their factories as machines and energy-source supplies change. To make money, they must make wise decisions at the right time, balancing environmental, business, and social concerns.

EVALUATION
Students are challenged to make money, but they cannot do that without considering all options and investigating the content of this simulation. The simulation will work best with groups of students entering information on a daily basis and discussing results, looking up information, and making decisions before the next day's turn. Students become actively involved in cooperative decisions—learning that to reach a group consensus, they must find information and share it. Students practice acquistional, organizational, creative, and communicative skills.

TITLE: EZ-CHEM GAS LAW SIMULATION

Company: CDL
Grade Level: 9–adult
Computer: Apple
Required Memory: 48K
Special Equipment: None

SUBJECT MATTER
Chemistry
Ideal gases and gas laws
Behavior of gases under varying conditions
Kinetic molecular theory

OBJECTIVES
- Visualize the behavior of gases under varying conditions
- Develop an understanding of the relationship between pressure, volume, and temperature through analysis of experimental data

PROGRAM DESCRIPTION
This program gives a graphical depiction of the relationship between pressure, volume, and temperature. The graphics let users try different temperatures, pressures, or volumes and watch what happens to the other variables. Students are encouraged to develop their own relationships through experimentation. They can discover the effect of temperature on volume, of pressure on volume, or of the number of moles on volume. They can stop the simulation to record data and then proceed. Data can be graphed and compared. Students can have a floating plunger or a controlled plunger. In the controlled-plunger experiments, they investigate ideal gases, diffusion, and mixing. They can also observe Brownian motion. They can use a $PV = nRT$ machine to obtain values on compressibility and have the computer gather and graph compressibility data for various real gases.

EVALUATION
The graphics in this program make it ideal for a whole-class demonstration. It is easy to move the plunger or change temperature or number of moles and collect data. This is also a good program for individual study. The program is inquiry-oriented and seeks to develop the intuitive understanding of how gases behave and how one can indeed develop a mathematical relationship among the variables.

TITLE: GEOLOGISTS AT WORK

Company: Sunburst
Grade Level: 7–adult
Computer: Apple, IBM
Required Memory: 64K (Apple), 256K (IBM)
Special Equipment: Printer very useful

SUBJECT MATTER
Geological field tests
Earth history
Geological processes
Scientific observation, hypothesis testing, record keeping

OBJECTIVES
- Investigate methods used in geological exploration
- Read, interpret, and construct geological cross sections, three-dimensional models, and geological maps
- Investigate the processes that have shaped the earth
- Use geological tools to analyze a land area to determine the processes that created it

PROGRAM DESCRIPTION
Students take core samples to see what rock layers are underground. From these samples, they create a geological cross section and suggest the history of the land mass. Geological maps and three-dimensional models can be created from the cross section. Students also hypothesize what natural resources might be present, using only a geological map and core samples.

EVALUATION
This is a companion to GEOLOGICAL HISTORY (which deals with cross sections and earth processes) and is more challenging. There are three different levels of problems, and students can create their own problems to challenge each other. Cross sections, maps, and three-dimensional models can be printed for reference. Students can use the software to investigate what different processes do to cross-sections and maps. The intellectual

processes that are required for the student to go from a two-dimensional cross section, to a three-dimensional model, to the extension on the land's surface (map) are difficult for students to visualize, and this program can be very helpful. The screen resolution makes the identification of some rock layers difficult; and it would be helpful to provide the correct geological explanations, rather than just a list of what happened.

TITLE: GEOLOGY IN ACTION

Company: Queue
Grade Level: 7–adult
Computer: Apple
Required Memory: 64K
Special Equipment: Printer useful

SUBJECT MATTER
Earth history
Geological processes
Geological cross sections

OBJECTIVES
- Visualize tectonic, depositional, and erosional processes
- Identify processes that shape the Earth's surface
- Interpret past geological events from cross sections
- Create geological cross sections that depict a particular geological history

PROGRAM DESCRIPTION
The program is divided into three parts: a tutorial, solving puzzles, and creating puzzles. The tutorial introduces students to geological concepts and processes that have formed the landscapes we see—and those that occurred millions of years ago. In the experimentation section, students try different geological processes to see what happens to the resultant cross section. The puzzle sections let students interpret graphical geological cross sections that are included on the disk and others made by the teacher or students. Students select processes and the graphical changes accordingly. They then compare the resultant cross section to a picture until the two match.

EVALUATION
This program, which is similar to GEOLOGICAL HISTORY, puts the learner in charge of the geological processes that shape the Earth. The tutorial explains the processes well, and the graphics depict what happens with each process. The experimentation and puzzle sections are inquiry-

oriented and allow students to try a process to see what happens to the Earth's surface. Puzzles can be saved and printed. The strategy of matching pictures is the process used by experienced learners, but it is seldom taught in the rush to have students "cover the material."

TITLE: GEOLOGY SEARCH

Company: McGraw-Hill
Grade Level: 6–12
Computer: Apple
Required Memory: 48K
Special Equipment: None

SUBJECT MATTER
Geology
Rock origins
Location of oil and natural gas
Scientific observation
Doing experiments
Testing hypotheses
Record keeping

OBJECTIVES
- Maintain the smooth and successful operation of an energy-exploration company
- Identify the advantages and disadvantages of different exploration techniques
- Make decisions based on content information and information available from the computer
- Read, record, and interpret data and make decisions based on that information
- Discover how to locate oil and natural gas

PROGRAM DESCRIPTION
Student teams explore the island of Newlandia in hope of finding oil, natural gas, and riches. They study the types of tests that would be conducted by geologists and can simulate density scans, core sampling, seismic blasts, and drilling. They must keep records of areas searched, results of tests, and money spent. Teams must research signs of oil and natural gas and decide what tests to run and when the results indicate the presence of oil.

EVALUATION

The students need to understand rock types, geological exploration techniques, and how to evaluate test results. The accompanying student manual provides a great deal of information. The teacher should know some likely sites in case prospecting is not successful early in the simulation. Students can have one turn a day, keeping track of their results. The program is interesting and challenging. GEOWORLD provides a similar challenge involving fifteen different resources on a worldwide basis.

TITLE: HARMONIC MOTION WORKSHOP

Company: High Technology Software Products
Grade Level: 10–adult
Computer: Apple, Atari
Required Memory: 48K (Apple), 32K (Atari)
Special Equipment: None

SUBJECT MATTER
Physics
Phase, amplitude, damping
Harmonic motion
Velocity, acceleration, kinetic and potential energy of objects in circular motion

OBJECTIVES
- Observe and compare objects executing simple harmonic motion
- Compare simple harmonic motion with corresponding uniform circular motion
- Observe and compare objects executing simple harmonic motion with those affected by a retarding force

PROGRAM DESCRIPTION
High-resolution graphics visually display simple and damped harmonic motion. An object is placed in harmonic motion on the screen, and students can alter phase, amplitude, and the damping factor. The graphic immediately shows the effects on motion. Instantaneous velocity and acceleration vectors, kinetic and potential energy values, and a corresponding object in circular motion can also be displayed. Other programs in the series are PROJECTILE MOTION, CHARGED PARTICLE WORKSHOP, and STANDING WAVE WORKSHOP.

EVALUATION
The graphics provide good visualization of all principles involved. The velocity and acceleration vectors are attached to the moving pendulum bob and provide a picture of the changes that occur. Potential and kinetic energy are shown in graphs alongside the moving bob, and the changes are

observed in real time. The relationships between simple harmonic motion and circular motion can be observed, and damping effects can be investigated. The program is particularly well adapted for use as a demonstration to the whole class but can also be used in small groups in an inquiry lab experiment.

TITLE: INTERACTIVE OPTICS

Company: EduTech
Grade Level: 7–12
Computer: Apple
Required Memory: 64K
Special Equipment: None

SUBJECT MATTER
Geometric optics
Lenses and mirrors
Fermat's principle
Snell's Law

OBJECTIVES
- Discover the relationships concerning lenses and mirrors
- Classify lenses
- Predict the path of beams of light passing through lenses
- Predict where objects can be seen in a mirror
- Have a practical hands-on experience with Fermat's principle

PROGRAM DESCRIPTION

The section on thin lenses lets students visualize and classify lenses, determine focal length, and determine where and what types of images can be "seen" for point sources and objects in both converging and diverging lenses. Students are asked to predict the path of beams of light prior to seeing how they actually are refracted. Images in mirrors are found through a hide-and-seek game, which students play with the computer. Fermat's principle is simulated with a lifeguard running on the beach and then swimming in the water to save another person. Students try to find the shortest time to reach the person by varying the amount of running and swimming time. Finally, the Snell's Law segment helps develop ideas about the index of refraction.

EVALUATION

The programs are hands-on and inquiry-oriented. Students are encouraged to predict and try prior to receiving factual information. The hide and seek with a mirror is an effective game, and Fermat's principle is well demonstrated with the swimming problem. These programs would be effective for demonstrating principles to the whole class or for groups of students working together on assignments. The entire set of programs provides explorations, concept introduction, and applications. The two "game" problems are stimulating.

TITLE: INVISIBLE BUGS

Company: MECC
Grade Level: 3–9
Computer: Apple II
Required Memory: 128K
Special Equipment: None

SUBJECT MATTER
Heredity and continuity
Dominant and recessive genes
Basic Mendelian genetics
Adaptation and natural selection

OBJECTIVES
- Control genetic variables to observe their effect on the physical appearance of offspring
- Observe, hypothesize, form, and test models
- Collect, organize, and interpret data
- Create a population of beetles whose physical appearances enable them to avoid detection by a predator

PROGRAM DESCRIPTION
The package consists of two programs, "Explore" and "Challenge." Students use "Explore" to develop an understanding of genetic relationships by collecting, organizing, and analyzing data. They can select sets of beetle parents, display the offspring from these crossings, and observe and compare the physical traits of the parents and offspring. Variables can be changed, and the offspring displayed as often as needed. Students may also elect to see the genes of the various bugs. The "Challenge" program requires students to demonstrate the understandings they have previously gained by creating a population of beetles with specific physical characteristics that will enable them to avoid detection by a predator. The program has easy, medium, and hard levels. To add to the challenge on the medium and hard levels, students must observe the interaction between a predator and a population of beetles to predict the set of traits that would best enable them to avoid detection by a predator.

EVALUATION

This program can be used with students from grades 3–9, to develop the basic concepts of Mendelian genetics and to practice a variety of scientific thinking skills. It has great depth in both content and features. Developmentally appropriate activities are suggested in the teacher's guide for students at four levels of understanding and intellectual maturity. Suggestions are also given for integrating the program with laboratory investigations in genetics. The core thinking skills from the ASCD framework are listed, and the applications that use them are identified. This program can be adapted very easily for use in a learning cycle format.

TITLE: IT'S A GAS

Company: Diversified Educational Enterprises
Grade Level: 9–adult
Computer: Apple
Required Memory: 48K
Special Equipment: None

SUBJECT MATTER
Chemistry
Ideal gases and gas laws
Behavior of gases under varying conditions
Mathematical solutions to gas law problems

OBJECTIVES
- Visualize the behavior of gases under varying conditions
- Solve ideal gas law problems
- Identify examples of the gas laws in action in everyday life

PROGRAM DESCRIPTION
The program is divided into five sections: algebra review, ideal gas laws demonstration, gas laws quiz, fire extinguisher simulation lab, and refrigeration simulation lab. The students can practice algebra problems, which require the same procedures and skills as the gas law problems. The program provides graphical demonstrations of gases as they behave under different pressure, temperature, and volume conditions. The "Gas Law Quiz" lets students test their knowledge of the gas laws, both conceptual and mathematical. A built-in calculator is available for working problems. After a visual explanation of how a refrigerator works, the refrigeration simulation challenges students to keep a refrigerator at an appropriate temperature for perishables. Students then apply what they have learned to the workings of fire extinguishers.

EVALUATION
The algebra review covers appropriate problems but does not allow students to do them in steps. The graphics throughout the programs are good, and students can get a good visual display of what is affecting gases.

The cause-and-effect relationships are very obvious in the graphical representations. Students can select independent and dependent variables for as long as they wish or quit when they have observed enough. The graphics are effective both for individual learning and for whole-class demonstration. This program can be part of the exploration, concept introduction, or application phases of the learning cycle. One problem with the program is that it uses non-standard symbols to represent recessive genes. This may cause some difficulty with the use of conventional symbols later in the learning process.

TITLE: LAWS OF MOTION

Company: EME
Grade Level: 7–adult
Computer: Apple
Required Memory: 48K
Special Equipment: None

SUBJECT MATTER
Physics
Motion
Doing experiments
Testing hypotheses

OBJECTIVES
- Develop skills of scientific observation
- State and test scientific hypotheses concerning motion
- Identify data that support or refute a given hypothesis
- Control variables in an experiment

PROGRAM DESCRIPTION
Students investigate motion in two worlds, where laws of motion are different. They observe simulated moving objects, predict what they will do, and analyze data in graphs as a block moves along a plane inclined at various angles.

EVALUATION
The format is inquiry-oriented, and students develop intuitive ideas about motion, possibly challenging misconceptions that they have developed. Students are challenged to try a variety of experiments and then to test the hypotheses that they have developed from the test results. They try to describe how these objects will behave and then can test to see if they are correct.

TITLE: MICROBE: THE ANATOMICAL ADVENTURE

Company: Synergistic Software
Grade Level: 9–12
Computer: Apple II
Required Memory: 48K
Special Equipment: None

SUBJECT MATTER
Human anatomy and physiology
Immunology
Effects of brain damage and other injury
Effects of drugs
Effects of bacterial, viral, parasitic, and fungal infections

OBJECTIVES
- Solve biological problems in a realistic setting
- Diagnose and apply treatment for medical problems
- Work cooperatively to complete a mission designed to solve a particular problem

PROGRAM DESCRIPTION
This is an educational simulation of the human body and its contents and the operation of a sophisticated research submarine. The scenario on which the game is based is similar to that of the movie "The Fantastic Voyage." The game can be played by one or several students. In the multiplayer mode, each student takes the role of one of the team: captain, navigator, physician, or technician. The game is begun by selecting a case from among those listed. All patients are in cryogenic state. After the case is accepted, a summary description of the medical problem is presented. The team then launches into the bloodstream of the patient to go to the source of the problem. A captain's manual, a navigator's chart, and a physician's reference are provided with the game. The master view screen can display a closed-circuit TV view outside the sub; a computer-generated map of the immediate surroundings; body maps; information displays such as blood samples, inventory lists, and library information; indicator lights; message windows; gauges; sonar; and compass readings. All com-

mands given and the status of the ship, crew, patient, and mission are displayed on the master view screen upon request.

EVALUATION

This game requires more than one player to be enjoyable and educationally profitable. It is an excellent example of a game that can be used with small-group cooperative learning methodology. It could probably generate a great deal of enthusiasm among senior high school life science students if used at the end of a detailed study of human anatomy and physiology to integrate learning from a variety of areas. Student involvement would be enhanced by having heterogeneous groups compete to cure the same patient. The potential for fostering constructive teamwork and scientific problem-solving skills is great with this program, but the instructor must determine that the time is well spent. It takes more than one class period to effectively complete one game.

TITLE: MINER'S CAVE

Company: MECC
Grade Level: 4–9
Computer: APPLE
Required Memory: 48K
Special Equipment: None

SUBJECT MATTER
Pulleys, levers, inclined planes
Wheels and axles
Mechanical advantage
Estimating, comparing, and inferring

OBJECTIVES
• Select appropriate machines based on their advantages
• Estimate the mechanical advantage of simple machines
• Use simple machines to complete a task

PROGRAM DESCRIPTION

Students can study four simple machines and learn about how each works. Mechanical advantages are estimated based on graphical representations. Students play a game to lift as much treasure as they can as quickly as possible. They must select the correct machine and adjust the load and mechanical advantage to raise the carts. Prior to the game, students can practice as much as they wish in an inquiry-oriented fashion to figure out how each machine works. They can adjust the length of lever arms, the size of a wheel and axle, the number of pulleys, and the length of a ramp to see if they can raise carts of different weights.

EVALUATION

This format is both challenging and fun. Students are encouraged to practice as much as possible in order to select the correct machine and adjust it quickly. If students select the wrong machine, they are penalized. If they do not adjust the machine properly, they are told why it will not work. Students will discover intuitively how these machines work and will be able to estimate advantage reasonably well.

TITLE: THE MBL PROJECT: MOTION

Company: Queue
Grade Level: 7–adult
Computer: Apple
Required Memory: 128K
Special Equipment: Red box, interface card, motion detector

SUBJECT MATTER
Speed, velocity, and acceleration
Time and distance graphs
Time and velocity graphs
Time and acceleration graphs
Scientific method

OBJECTIVES
- Use scientific instruments to gather data on real-world phenomena
- Perform scientific experiments on familiar objects
- Distinguish between distance, velocity, and acceleration
- Interpret bar graphs and line graphs
- Measure velocity and acceleration

PROGRAM DESCRIPTION
The motion detector sends out and receives back a signal and uses this to determine the distance from the detector. The program uses this information to provide distance, velocity, or acceleration information. The graphs can be seen in real time, as the object moves, so that students can begin to develop ideas about what a particular graph represents. Students have control of the graphs and can recall and overlay data. The motion can come from a toy car, a real car, a student moving toward or away from the detector, or any moving object. The motion detector can be held out the window, facing down toward the ground. An object dropped from beneath the detector will be recorded by the computer, and the acceleration due to gravity can be explored.

EVALUATION

The hardware requires installation inside the computer; but once this is done, it is easy to use. This program is designed to encourage experimentation with motion and to test hypotheses. The menu is clear, and students can save and recall information. This MBL is an excellent aid in motivating students and encouraging them to run many different experiments. Students can be challenged to produce a graph with a specific appearance or to interpret what they see on the screen.

TITLE: MYSTERY MATTER

Company: MECC
Grade Level: 3–9
Computer: APPLE II
Required Memory: 128K
Special Equipment: None

SUBJECT MATTER
Properties of matter
States of matter
Solids, liquids, gases

OBJECTIVES
- Identify matter according to its chemical and physical properties
- Correlate the testing tools with the chemical properties they test
- Preplan and control the creation of a substance
- Record, organize, and evaluate information to solve a problem

PROGRAM DESCRIPTION
This is a discovery-learning simulation that allows students to use a variety of tools to test unknown matter for designated physical and chemical characteristics. Students then use the information they gather through the tests to identify the mystery matter.

EVALUATION
This program can be used very effectively in conjunction with concrete activities in the primary and intermediate elementary grades to develop skill in observation, classification, prediction, and making inferences. The program can be used flexibly in two modes. In "Matter Search," students use the simulated robot lab tester to identify the mystery matter hidden among several different substances. This mode can be used to either introduce or follow up "real" mystery-substance activities. In the "Matter Maker" mode, students can access or add to a database of common chemicals. This is a very useful feature that could also be used to complement or extend concrete activities.

TITLE: MYSTERY OBJECTS

Company: MECC
Grade Level: 2–4
Computer: Apple II
Required Memory: 128K
Special Equipment: None

SUBJECT MATTER
Physical properties of matter

OBJECTIVES
- Observe and compare objects according to such physical properties as shape, color, size, weight, texture, and smell
- Determining the identity of unknown objects from descriptions of their physical properties
- Recording, organizing, and evaluating information to solve a problem

PROGRAM DESCRIPTION
In "Mystery Objects," students are challenged to determine the identity of an unknown object by testing it for various physical properties, using animated "data snoopers." As the unknown object is tested for its properties, students observe and record the results. This information enables them to identify their mystery object.

EVALUATION
This program is flexible in design and can be used in a variety of ways to promote scientific problem-solving skills. The program can be used in all phases of the learning cycle. Groups of young children will find it easy to manipulate, and it works well as a teacher demonstration during concept introduction. Two levels of interaction are available. In the "Practice Sessions," the skill level required is lower, and hints are given. The practice sessions, with the aid of the teacher, can help students in developing successful problem-solving strategies. In the "On Your Own" section of the program, no hints are given, and the problems require higher-level skill. This program could easily be integrated with real mystery box and mystery bag laboratory experiences.

TITLE: NICHE

Company: Diversified Educational Enterprises
Grade Level: 8–12
Computer: Apple II
Required Memory: 48K
Special Equipment: None

SUBJECT MATTER
Ecological niche
Population
Competitors
Habitat
Predators

OBJECTIVES
- Develop an understanding of ecological niches by actually attempting to place organisms in their proper niches
- Learn about interrelationships in ecosystems through manipulation of ecological variables
- Improve problem-solving skills

PROGRAM DESCRIPTION
This is an interactive program, which uses a game format to explore the concept of an ecological niche. Students must attempt to correctly place one of five organisms into its proper niche by specifying the environment, range, and competitors for the organism. The organism may flourish in a well-specified niche or fail in a poorly specified one. At the conclusion of each turn, the program reports the amount of food present in the range, the amount of food taken by competitors, the amount of food available to the population of the selected organism, deaths due to predation, deaths due to starvation or other causes, births, and the current population size. If students specify the organism's niche well, there will be few deaths, many births, and the population will grow. If they specify the niche poorly, the population will decline. Barring disaster, the game continues for five turns.

EVALUATION

This program is a simple, enjoyable game that helps students develop problem-solving skills as they develop a qualitative understanding of the niche concept. This program is intended to approximate organism/environment agreement, competitors, and true ranges related to the organism selected. The game could be used effectively in the exploration and concept introduction phases of the learning cycle, as the concept of niche is being formed.

TITLE: ODELL LAKE

Company: MECC
Grade Level: 4–6
Computer: Apple II
Required Memory: 64K
Special Equipment: Color monitor recommended

SUBJECT MATTER
Organisms in a freshwater aquatic environment
Aquatic food chains and food webs
Habitats of aquatic organisms
Predator/prey relationships

OBJECTIVES
- Determine the predator/prey relationships among a group of twelve organisms in a specific biological community
- Create a diagram of the food web of the community based on the list of the predator/prey relationships
- Fit these specific relationships into a general conceptual framework of the relationships among living things and apply the standard terms to these concepts
- Apply these general concepts to a variety of specific biological communities with which the students are familiar

PROGRAM DESCRIPTION
This is a game where students role play the fish in the simulated Odell Lake. The simulated fish encounter other creatures and collect information about their relationship to those creatures. In "Go Exploring," students choose the fish they wish to role play. They can use the ? key to access a screen that will give them information about the fish. After students choose a fish, they have eleven encounters with other organisms. With each encounter, they must select one of five possible responses (eat, chase away, ignore, shallow escape, or deep escape). The game ends if the students are eaten by another animal or if they successfully survive all eleven encounters. In "Play for Points," the opposing organism is shown, but not identified. Students have a limited amount of time to react to each opposing organism. Points are awarded for the quality of the choice. As the game progresses, the

computer chooses the role the students must play, and the amount of time allowed for the choice diminishes. The game continues until the students are eaten by another animal. The score represents the students' ability to quickly identify the relationship between the organisms in Odell Lake.

EVALUATION

ODELL LAKE is a highly motivating learning tool for elementary life science. "Go Exploring" can be easily adapted for use during the exploration phase of the learning cycle, providing students with a great deal of control over the learning process. The teacher can use this section for the concept introduction phase as well. The game is ideal for whole-class demonstration of concepts during that phase. The "Play for Points" section can be used to good advantage in the concept application phase, with groups of students competing with each other for points. The teacher's guide is very helpful, with suggestions for use, discussion topics, follow-up activities, and student worksheets. A discussion of the Odell Lake model along with its limitations is very complete. This is an excellent program that students with a variety of learning styles will enjoy.

TITLE: PLAYING WITH SCIENCE: TEMPERATURE (MBL)

Company: Sunburst
Grade Level: K–9
Computer: Apple
Required Memory: 64K
Special Equipment: Cables and probes

SUBJECT MATTER
Temperature and heat
Identifying hypotheses
Identifying variables
Performing scientific experiments
Analyzing tabular and graphical data

OBJECTIVES
- Use scientific instruments to gather data on real-world phenomena
- Perform scientific experiments on familiar objects
- Distinguish between thermometers and temperature
- Distinguish between heat and temperature
- Interpret bar graphs and line graphs
- Describe and predict temperature variations

PROGRAM DESCRIPTION
The three temperature probes (thermistors) plug into the game port and permit students to measure temperature of three different objects simultaneously. Students can run experiments for various lengths of time, recording, analyzing, and graphing results. Temperatures can be presented with a digital readout or on a graph; and students can change the type of graph, the temperature scale, and other variables.

EVALUATION
Well-designed hardware and software make it easy for students to conduct experiments. The hardware is durable and easy to install and use. The three different probes permit a wide range of experiments to be performed, and the teacher's guide is extensive, containing lesson plans, student activity sheets, and setup information.

TITLE: POLLUTE

Company: Diversified Educational Enterprises
Grade Level: 9–12
Computer: Apple II, IBM
Required Memory: 64K
Special Equipment: None

SUBJECT MATTER
Pollution of aquatic ecosystems
Waste treatment
Biological oxygen demand

OBJECTIVES
- Develop skill in data interpretation and evaluation
- Develop investigative abilities and experimental skills through manipulation of variables
- Sharpen decision-making skills by applying them to an ecological problem
- Develop a model for the interaction between the living and nonliving world in an aquatic ecosystem
- Develop a realistic framework for environmental decision making by identifying and working within economic, social, and ecological constraints

PROGRAM DESCRIPTION
This student-interactive simulation studies the impact of pollutants on the oxygen content and fish life of various bodies of water. Students study the impact of water temperature, waste type, waste treatment, rate of waste dumping, and type of body of water on the oxygen content and survival of fish. The computer generates graphical and tabular data to enable students to test various hypotheses. The teacher's manual contains background on the simulation, suggestions for classroom use, a student laboratory manual, and a problem-solving test keyed to the objectives of the program.

EVALUATION

This program effectively combines a laboratory on biological oxygen demand (BOD) with a computer simulation to promote the development of a model for the pollution of an aquatic ecosystem, the development of scientific problem-solving skills such as making inferences and predictions, and formulating strategies based on a consideration of both economic and environmental factors. The problem exercises in the student laboratory manual are carefully written to involve students in decision making that incorporates more than one perspective.

TITLE: PROJECT SERAPHIM

Company: University of Wisconsin
Grade Level: 9–adult
Computer: Apple, IBM, Atari, Commodore, Macintosh, TRS-80
Required Memory: Varies
Special Equipment: Varies

SUBJECT MATTER
All areas of general high school and college chemistry
Environmental/industrial chemistry
Programmer's tools
Statistical analysis
Interfacing

OBJECTIVES
- The programs provide tutorials, drill and practice, simulations, laboratory modules, databases, and other information involving all aspects of chemistry.

PROGRAM DESCRIPTION
Project SERAPHIM began in 1982 as an NSF-sponsored clearinghouse for instructional microcomputer information in chemistry. It not only collects and distributes materials and chemistry software but also offers teacher training, research, and development. Project SERAPHIM has over 600 microcomputer programs on about 175 different disks.

EVALUATION
Project SERAPHIM assigns its software to one of three categories. Category I software is fully reviewed and user-tested; corrections and changes have been made to these programs. Category II material has not been fully reviewed but has been distributed for at least one year as a Category III program with no negative comments. These programs run reasonably well and are generally accurate. Category III programs are untested and may contain bugs or errors. These may be top-quality programs but have not yet been fully reviewed. Programs and interfacing modules are very reasonably priced and usually well documented.

TITLE: PROJECTILES II

Company: Vernier
Grade Level: 9–adult
Computer: Apple
Required Memory: 64K
Special Equipment: None

SUBJECT MATTER
Physics
Projectile motion
Scientific observation
Hypothesis testing

OBJECTIVES
- Observe the motion of projectiles and predict their landing points
- Investigate the variables involved in projectile motion
- Gather data and describe the relationships between launch speed, launch angle, velocity, and distance

PROGRAM DESCRIPTION
Nine challenges are presented to provide an opportunity to experiment with the variables that affect projectile motion. Projectiles are launched from a cliff or from the ground at a randomly placed or a moving target. The projectile is given an initial velocity and a launch angle. There may be air or wind resistance, or there may be no resistance. Users fire at the target until the target is hit by the projectile. Two challenges are provided, which encourage exploration of the relationship between two variables.

EVALUATION
These challenges maintain interest, and the projectile motion track can be left on the screen for future reference. Problems can be solved by trial and error, by mathematical calculations, or by estimation based on observation. Both intuitive and mathematical relationships can be explored, and the software encourages students to actually explore the relationships. The program can be used effectively as a demonstration for the whole class, with the class suggesting launch variables. Explanations are not provided on the screen, but careful record keeping allows users to infer the relationships. The formulas and solutions for mathematical calculations are included and explained in the manual.

TITLE: RAY TRACER

Company: Vernier
Grade Level: 9–adult
Computer: Apple
Required Memory: 64K
Special Equipment: None

SUBJECT MATTER
Physics
Geometric optics
Index of refraction
Scientific observation and prediction

OBJECTIVES
- Demonstrate the concepts of reflection and refraction
- Gain insights into how light behaves
- Observe and predict the refraction of light rays as they pass from one medium to another

PROGRAM DESCRIPTION
Users choose media such as air/glass, diamond/glass, or any combination. They also select whether to use a straight interface, a lens, a mirror, or a raindrop. The interface and the devices are then drawn on the screen, and a starting point and a target are randomly selected. Users are then asked to enter a starting angle, and the ray is traced on the screen. Users enter a new angle, and another ray is drawn, until the target is hit. The angle of entry and the angle of the refracted/reflected ray are given on the screen.

EVALUATION
The graphics are colorful, and the behavior of the light rays is easy to observe. (A few of the programs contain the calculated angles only and do not contain graphics.) The program can be played as a game, with students trying to see how few trials it takes to hit the target. The program also would work well as a classroom demonstration during the discussion of geometric optics. These light rays are easier to see than those from a light box, and each trial remains on the screen for comparison until erased. The directions for use of the program are in the manual, and there is little on-screen instruction.

TITLE: ROCKY: THE MINERAL IDENTIFICATION PROGRAM

Company: Michigan State University
Grade Level: 7–adult
Computer: IBM PC or CDC mainframe
Required Memory: 64K (IBM)
Special Equipment: Three sets of minerals, hardness kit, streak plate

SUBJECT MATTER
Geology
Mineral identification
Scientific observation
Testing hypotheses

OBJECTIVES
- Evaluate the physical properties of minerals
- Develop a process of discriminating mineral properties
- Use properties of minerals to identify hand specimens

PROGRAM DESCRIPTION

This program provides information on five physical properties—luster, shape, color, specific gravity, and hardness—for fifty-eight representative minerals divided into three sets. The program allows students to evaluate the physical properties in any order and to learn a best strategy for discriminating among the minerals in the set. Students select one mineral and enter any identified property. A list of the minerals in that set, which have that property, is then presented on the screen. Students can then enter additional properties, eliminating more minerals. This process continues until only one mineral name remains. The microcomputer version is menu-driven. The mainframe version contains a natural language dialogue.

EVALUATION

This program not only teaches how to do the tests and how to identify minerals but, more important, it also teaches discrimination and critical attributes. The program is free-choice and not limited by branching routines that do not fit the learner. In addition, students do not have to follow the same routine in the same order, and they can develop their own best strategies for identification. The minerals in the sets are designed both for secondary and college use and contain some interesting specimens. This program is based on sound learning strategies and shows science as a process.

TITLE: SCI-LAB

Company: Sargent-Welch
Grade Level: 9–adult
Computer: Apple
Required Memory: 64K
Special Equipment: I/O key, A/D card, converter box, probes

SUBJECT MATTER
Biology
Chemistry
Physics
Scientific experimentation

OBJECTIVES
- Become familiar with biology, chemistry, and physics concepts
- Understand how to set up and calibrate scientific apparatus
- Use scientific instruments to gather data on real-world phenomena
- Perform scientific experiments on familiar objects to discover how the world operates
- Develop and test hypotheses
- Interpret bar graphs and line graphs
- Interpret data collected by the computer

PROGRAM DESCRIPTION
The MBL software exists in the following areas:

- Biology: Electrical activity of the heart, evaluation of physical conditioning, mechanics of respiration, rate of photosynthesis, rate of yeast fermentation, and skeletal muscle activity
- Chemistry: Energy and phase changes, heat of fusion, hydronium ion concentration and pH, indicator colors and pH, titration of a strong acid, and measuring temperature and heat
- Physics: Acceleration, acceleration due to gravity, average and instantaneous velocity, force of friction, measurement and metric system, and resolution of forces

EVALUATION

Each experiment is a separate package, with all the interfacing equipment and probes needed. The additional scientific equipment is clearly identified, and each program has a tutorial on the main concepts of the experiment. The directions for each experiment are very clear and presented in high-resolution graphics. Many of the tutorials also contain high-resolution graphical models of the concepts. The data gathered by the computer can be used individually, or group data can be gathered. The program can be used as a prelab or postlab or in place of lab experiments that cannot otherwise be performed.

TITLE: SCIENCE TOOLKIT: MASTER MODULE (ALSO MODULE 1: SPEED AND MOTION; MODULE 2: EARTHQUAKE LAB; MODULE 3: BODY LAB)

Company: Broderbund
Grade Level: 4–12
Computer: Apple
Required Memory: 64K
Special Equipment: Interface box, cables, probes

SUBJECT MATTER
Heat and temperature
Light intensity
Velocity and acceleration
How to measure earthquake intensity
Heat rate
Lung capacity

OBJECTIVES
- Use scientific instruments to gather data on real-world phenomena
- Perform scientific experiments on familiar objects
- Distinguish between heat and temperature
- Interpret bar graphs and line graphs
- Measure velocity
- Investigate variables associated with the motion of a balloon-powered car
- Measure heart rate and lung capacity
- Model the behavior of a seismoscope

PROGRAM DESCRIPTION
Broderbund produces school editions containing features not found in the versions sold to the general public. The MASTER MODULE is necessary to run all other modules but does not contain a printing option. The hardware connects through the game port, and the probes are plugged into a small black box. The program is menu-driven and easy to use. The methods of operation are explained in the guide, and suggested experiments are relevant to most science curricula.

EVALUATION

The nonschool edition will not measure temperatures above 60°C, and the program will sound a warning that the experiment is too hot. The hardware is easy to use, but the cables are short and limit the use of some of the probes. An extension cable is provided in the SPEED AND MOTION module, and several more would be helpful. Students are encouraged to keep good records, and experiments are inquiry-oriented.

TITLE: SIMULATIONS IN EARTH SCIENCE

Company: Data Command
Grade Level: 3–9
Computer: Apple
Required Memory: 48K
Special Equipment: None

SUBJECT MATTER
Earthquakes
Volcanoes
Weather
Using instruments to extend the senses
Application of concepts

OBJECTIVES
- Visualize earthquakes, volcanoes, and weather phenomena
- Use simulated instruments to aid in measurement and prediction of Earth science phenomena
- Relate observations from instruments or observations to actual and predicted events
- Learn vocabulary and concepts associated with earthquakes, volcanoes, and weather

PROGRAM DESCRIPTION
This series of four disks covers several intermediate concepts in a simulation format. The "Earthquake" and "Volcanoes" programs simulate Earth movements, causes, and results. Instruments are used to measure phenomena and predict earthquakes and weather phenomena. The "Weather Instruments" program demonstrates information about the use of instruments and provides experience in recording data, using the English system of measurement. The "Clouds" simulation provides instruction about the different types of clouds and offers a simulated airplane flight around and through various clouds. The student management system lets students leave the program and begin where they left off. The program includes review questions, tutorials, simulations, and posttests.

EVALUATION

These programs come with a teacher's guide that integrates the software with an entire unit on each subject. The programs are short and well designed. Some of the definitions are too simplistic, and some of the weather instruments need further explanation. The tutorials provide explanations, and the application phases explain incorrect responses. Feedback is immediate and appropriate, and students may try missed questions again.

TITLE: SIR ISAAC NEWTON'S GAMES

Company: Sunburst
Grade Level: 3–12
Computer: Apple; IBM PC, PC JR, and PS/2; Tandy 1000; Commodore
Required Memory: Apple, 48K; IBM PC and PC JR, 128K; IBM PS/2,
    256K; Tandy 1000, 256K; Commodore, 64K
Special Equipment: None

SUBJECT MATTER
Motion
Forces
Momentum
Frictions
Testing hypotheses
Building models

OBJECTIVES
- Gain an intuitive understanding of the relationships among velocity, change in velocity, and applied force
- Gain an intuitive understanding of frictional forces
- Compare a friction-free environment with that of Earth
- Develop strategies for determining position and velocity before applying force

PROGRAM DESCRIPTION
These games can be played way out in space (frictionless), on Earth, or close to the sun. The program is a microworld in which a marker is moved around a track without crashing into the walls. Students must guide the marker by kicking it in a certain direction and watching it as it moves, coasts, or stops. This movement can occur on sand, grass, or ice in any one of the three environments. In another game, students kick the marker around obstacles to a finish box. Two students can compete by trying to race each other to the box, or one can try to catch the other.

EVALUATION

The nature of motion is a concept that alludes most adults, who do not understand Newton's concepts of motion or more recent theories. To understand these deep concepts, students must develop intuitive ideas that reflect how things really move. These games help promote such qualitative understandings. In these games, time stops so that students can consider the next move, while in reality the marker is still moving. This must be accounted for when the marker is kicked again. Students can investigate many "what if" questions.

TITLE: TELLSTAR—HALLEY'S COMET EDITION

Company: Spectrum HoloByte
Grade Level: 6–adult
Computer: Apple II, IBM, Macintosh
Required Memory: Varies
Special Equipment: None

SUBJECT MATTER
Astronomy
Stars, constellations, near- and deep-sky objects
Apparent and real movement of stars
Location of objects in the night sky

OBJECTIVES
- Visualize the night sky
- Observe patterns and identify objects and constellations
- Compare the nighttime sky at different times of the year
- Compare the nighttime sky at different times in history
- Compare the nighttime sky from different locations on Earth.
- Locate objects in the real sky

PROGRAM DESCRIPTION
The program puts a planetarium on the computer screen. Users can specify the viewing location, time, and date. The software can save one view at a time, and a specified view can be recalled for later viewing. The program displays both northern and southern hemispheres and includes Messier objects. Students can save screen dumps to a data disk and print them with a utility program. The program adjusts for precession and permits observation to any azimuth along the horizon or straight overhead. Students can enter the name of an object, and the computer will find it. Information concerning right ascension, declination, phases, rising, setting, and the like is also available.

EVALUATION
The program may take up to four minutes to compute all the stars, constellations, planets, and objects once it is given a new location. This is a

good program for people interested in learning about the skies or for finding objects. The HALLEY'S COMET EDITION also plots the path of the last approach of that comet. Although there are more precise programs for the amateur and professional astronomer, this is a good program for beginners and amateurs who do not require greater precision. The ability to move to any azimuth quickly is almost like being outdoors and turning around to look at the sky.

TITLE: TIMELINER

Company: Tom Snyder Productions
Grade Level: K–12
Computer: Apple
Required Memory: 64K
Special Equipment: Printer

SUBJECT MATTER
Timelines—any subject, any complexity, any grade level

OBJECTIVES
- Develop a perspective of the time and of the relationship among historical events
- Discover patterns in history
- Develop hypotheses about the relationships among events
- Understand personal history in relationship to other events
- Research and organize information

PROGRAM DESCRIPTION
The program comes with data disks for "Science and Technology" as well as for various historical periods. The class can create a timeline for a topic they are studying, change the scale, print it out, and see "what happened when." Separate timelines can be merged to see patterns or relationships. Timelines can be viewed on the screen prior to printing.

EVALUATION
This program is a valuable tool in helping students visualize changes over time. Problems of the vast amount of geological time or the relationship among inventions leading to the Industrial Revolution can easily be seen. Preparing personal timelines and comparing them helps students begin to get a perspective on time.

TITLE: VOYAGE OF THE MIMI

Company: Sunburst
Grade Level: 6–12
Computer: Apple II
Required Memory: 128K
Special Equipment: VCR (for the video segments)

SUBJECT MATTER
Aquatic ecosystems
Whales and their environment
Mapping and navigation
Computer literacy

OBJECTIVES
- Examine the effect of human activities on selected plant and animal species
- Use simulated instruments and maps to locate and free a whale caught in the net of a fishing trawler
- Write simple programs in "turtle graphics" to create graphic images on the computer
- Use an MBL kit to conduct experiments with heat and temperature, sound and light, and apply their findings to the effects of these physical phenomena on whales and people

PROGRAM DESCRIPTION
This package uses a multimedia, multidisciplinary approach to teach an integrated set of concepts in science, math, social studies, and language arts. The package includes four computer programs, twenty-six fifteen-minute video segments, maps, and other printed materials. The video episodes follow a research expedition studying the humpback whale off the coast of New England. The story serves as a backdrop that provides an accurate description of how scientists think and presents problem situations to stimulate student thought. Besides the basic story line, the video segments provide informational visits to museums and other resources.

The "Ecosystems with Island Survivors" disk introduces students to the elements of and the relationships within ecosystems, with emphasis on food chains and food webs. In this simulation, students select plant and animal species to inhabit two ecosystems and study how they are affected by human activity. The "Maps and Navigation" disk helps students integrate the two-dimensional world of maps with the three-dimensional world of navigation. It provides a simulation in which students take the role of a ship's crew to use its instruments to locate a whale trapped in the net of a fishing trawler. In the context of navigational games and simulations, the "Introduction to Computing" disk introduces "turtle graphics" and some higher-level computer concepts. The "Whales and Their Environment" disk provides a simulated laboratory in which students gather and display data about the real world. Students conduct simulated experiments on how physical phenomena affect whales and people.

EVALUATION

The program is a well-integrated set of multimedia materials designed to teach important ecological concepts as well as to offer the opportunity for students to apply scientific problem-solving skills to real-life problems. It can be easily adapted for middle schools with curricula that emphasize the interconnections between math, science, social studies, and language arts. The video segments and the computer materials are likely to catch and hold the attention of young adolescents. A sequel called THE SECOND VOYAGE OF THE MIMI is now available. In the second voyage, students are taken on an adventure to the Yucatan Peninsula of Mexico to study the ancient Maya civilization. Computer programs on mathematics and the astronomy of the Earth and sun are included in the second package.

TITLE: WEATHER FRONTS

Company: Diversified Educational Enterprises
Grade Level: 4–12
Computer: Apple, TRS-80
Required Memory: 48K (Apple), 32K (TRS-80)
Special Equipment: None

SUBJECT MATTER
Warm and cold fronts
Weather symbols and weather maps
Weather forecasting

OBJECTIVES
- Identify weather symbols
- Interpret a weather map
- Identify high- and low-pressure systems from wind movement
- Prepare a weather map from data provided
- Predict weather from past weather data

PROGRAM DESCRIPTION
Students study weather symbols and weather maps from supplementary material provided. The program presents four different weather situations: a winter cold front, a winter warm front, a summer cold front, and a summer warm front. A graphic shows the southern Lake Michigan area, and the daily temperature, wind speed and direction, barometric pressure, dew point, cloud cover, and weather are printed on the screen. The weather is given for four days, and students must predict it for the fifth day.

EVALUATION
The package includes a clear manual designed to teach weather symbols, weather maps, and weather prediction; this manual must be closely integrated with the software for maximum effect. The graphics are good but sometimes superfluous to the assignments. They do keep attention focused and look like a newspaper weather map. Students can then go to the newspaper and apply the skills learned in the program. Students are asked to predict each weather feature separately and are evaluated immediately. If the answer is wrong, the software explains the reasons for the right answer. The program can be an excellent application or extension to a lesson.

TITLE: WOOD CAR RALLY

Company: MECC
Grade Level: 3–9
Computer: Apple
Required Memory: 128K
Special Equipment: Color monitor suggested

SUBJECT MATTER
Motion and the laws of motion
Forces
Gravity
Friction

OBJECTIVES
- Control five variables in order to observe their effect on the distance a wood car will travel
- Observe, hypothesize, form, and test models
- Design experiments
- Collect, organize, and interpret data to solve a problem

PROGRAM DESCRIPTION
This is a discovery-learning simulation that uses a "wood car rally" to challenge students to determine how each of three car characteristics and two ramp characteristics affect the distance a wood car will travel down a runway.

EVALUATION
This program has two major strengths. Besides providing an interesting and challenging physical science learning environment for upper elementary and middle school students, it can be flexibly used in any phase of the learning cycle. The "Practice Track" section, which is unstructured and unguided, can be used in the exploratory and concept introduction phases in a whole-class or small-group setting. The "Competition Track" section can be used to good advantage in the application phase of the learning cycle by having groups of students compete against each other. The program can also be used to support and complement actual hands-on experiments in the classroom. Students can use the program to guide them in the building of real Pinewood Derby cars.

GLOSSARY OF IMPORTANT TERMS

Academic Learning Time (ALT). The amount of time a student spends attending to relevant academic tasks while performing those tasks at a high rate of success.

ALT. See *Academic Learning Time.*

Artificial Intelligence. The use of computers to imitate or expand human intelligence. Computers that play chess usually employ artificial intelligence. Another example is found in "expert systems" being developed to offer physicians "second opinions" on their diagnoses of patients.

Authoring Language. A computer program that lets the user enter commands that instruct the computer to carry out various tasks, such as the presentation of a drill or tutorial. An authoring language is usually easier to learn but less flexible than a programming language.

Auxiliary Storage Device. Any device (such as a tape or a floppy disk) on which programs and other computer data can be stored in order to be transferred into the computer's memory.

BASIC. One of the most commonly used programming languages for microcomputers.

Branching Programmed Instruction. A form of programmed instruction in which the nature of each step in a learning sequence is determined by the student's response at the previous step.

CAI. See *Computer-Assisted Instruction.*

Cathode-Ray Tube (CRT). The "television" screen that displays output from the computer.

CBE. See *Computer-Based Education.*

CMI. See *Computer-Managed Instruction.*

Computer-Assisted Instruction (CAI). The use of the computer to provide instruction directly to the learner. When students run a drill, tutorial, or simulation, they are engaged in CAI.

Computer-Based Education (CBE). A synonym for computer-managed instruction.

Computer-Managed Instruction (CMI). The use of the computer to coordinate instructional activities. CMI can be used in conjunction with CAI, but it can also be used to coordinate noncomputerized modes of instruction.

Computer Program. A set of instructions that make the computer carry out specified operations.

Conjectural Use of Computers. The use of computers to build interactive models of various phenomena so that hypotheses or configurations can be formulated and tested by the learner. Examples include microworlds and model-building situations.

Corrective Feedback. Information that explains to the learner the nature of a mistake or suggests ways to move from the incorrect answer to a correct answer.

Courseware. Computer software designed for instructional purposes.

CRT. See *Cathode-Ray Tube.*

Cursor. The "prompt" (often a flashing box or a flashing line) that indicates where the next entry will take place on the CRT screen.

Database. An organized set of information. A database management program provides electronic access to a set of information by permitting entry, storage, sorting, and retrieval of data.

Direct Instruction. Academically focused, teacher-directed interactions using sequenced instructional materials.

Disk Drive. A mechanism into which floppy disks are inserted in order to transfer information to and from the computer's random access memory (RAM).

Documentation. Hardcopy or electronic information that describes how to use a piece of hardware or software.

Drill. A program that provides repeated practice and feedback regarding a skill or concept.

Electronic Bulletin Board. An electronic communication system, usually accessed via a modem, which enables users to share information about topics of common interest.

Emancipatory Use of Computers. The use of computers to free learners from routine, nonproductive work—that is, from work that does not contribute to the lesson objectives.

Feedback. Information indicating that a response is right or wrong. Feedback may be either positive or negative.

Floppy Disk. A small and compact auxiliary storage device on which information can be kept for subsequent transfer to the computer's random access memory (RAM).

Gradebook. A program designed to keep records and generate reports regarding student performance.

Graphics. Diagrams and pictures drawn with the aid of the computer.

Hardcopy. Output that is printed on a permanent surface (such as paper) instead of merely appearing on a temporary surface (such as a CRT screen).

Hardware. The physical equipment that comprises a computer system. It is differentiated from the software, which runs on this physical equipment.

Help Screen. A screenful of information that provides instructions on a specific topic or answers to a question that the learner asks. Help screens usually appear in response to a request initiated by the learner.

ICAI. See *Intelligent Computer-Assisted Instruction.*

Input. Information that is sent into the computer's memory, usually from a keyboard, from an auxiliary storage device, or through a modem.

Instructional Use of Computers. The use of the computer to provide instruction directly to the learner. Examples include drills and tutorials.

Intelligent Computer-Assisted Instruction (ICAI). The application of principles of artificial intelligence to computer-assisted instruction that enables the computer to analyze learner characteristics and to adjust its presentation of information in response to these characteristics.

Interactive Videodisc System. A combination of a computer with a videodisc that displays visual and auditory sequences in response to input provided by the learner.

Interface Card. An electronic connection device that fits inside the computer to add additional elements (such as disk drives, printers, modems, and MBL devices) to a computer system.

Joystick. An instrument that permits the user to move the cursor or in some other way control movement on the CRT screen.

Keyboard. The "typewriter" portion of the computer, which permits entry of various characters into the computer's memory.

Kilobyte (K). One thousand bytes. The computer's memory size is normally indicated in kilobytes.

Linear Programmed Instruction. A form of programmed teaching in which all learners go through the lesson from beginning to end, in exactly the same sequence.

MBL. See *Microcomputer-Based Laboratory.*

Memory Expansion Card. A device, inserted into the computer, that contains memory chips and increases the random access memory (RAM) of the computer.

Memory Size. The number of bytes of random access memory (RAM) that the computer makes available to the user. In general, computers with greater memory sizes can run more complex programs and store larger amounts of data than those with smaller memories.

Menu. A screenful of information and prompts that enable the learner to choose from a list of activities that the computer can perform.

Microcomputer. A relatively small computer that employs a microprocessor. Microcomputers are also referred to as personal computers. They are smaller than minicomputers, which in turn are smaller than mainframe computers.

Microcomputer-Based Laboratory (MBL). An application of the microcomputer in which the computer is connected (interfaced) directly with a science experiment to automatically perform such tasks as taking measurements and collecting, tabulating, and graphing data.

Microworld. A computerized setting in which the rules of the "world" are defined by the learner or by the programmer to enable the learner to test hypotheses and develop an intuitive understanding of that world.

Modem. A device for transferring information from one computer to another, usually across telephone lines.

Monitor. A CRT screen that is connected to a computer system and displays the input going into the system and the output coming from it.

Monochrome Monitor. A monitor that displays output in a single color on a background of another single color (e.g., black on white). It is distinguished from a color monitor, which will generate a wide variety of background and foreground colors.

Mouse. A device for moving the cursor or in some other way controlling movement on the CRT screen. A mouse differs from a joystick in that it controls movement more precisely but is less suited to many gamelike activities.

Negative Feedback. Information indicating that the learner's response was incorrect. If negative feedback also explains to the learner the nature of the mistake, it becomes corrective feedback.

Network. A series of computers that are connected in some way. Some networks are permanent (the computers always work together), whereas others are temporary (the computers can be joined together for particular operations).

Output. The information provided by the computer in response to instructions or input from a user. The output usually appears on the CRT or on a sheet of paper, but it may be sent directly to an auxiliary storage device or to another computer.

Peripheral Devices. Any of the devices that are added on to the main computer system. Peripherals include disk drives, printers, joysticks, modems, and so on.

Port. An external point of connection for the peripheral devices associated with a computer system, such as a printer or modem. Ports often come built into the computer and serve the same functions as interface cards, which often must be added to the system.

Positive Feedback. Information indicating that the learner gave the correct response. Positive feedback is usually considered to be a form of reinforcement.

Printer. A mechanism for generating hardcopy, printed output from a computer.

Probeware. A hardware and software package that enables the learner to use the computer as a laboratory by taking measurements through instruments that are directly interfaced to the computer.

Programming Language. A set of rules and commands that instruct the computer to carry out various tasks. A programming language is usually more difficult to learn but also more flexible than an authoring language.

Prompt. This term has two separate meanings in instructional computing. First, the cursor or some other symbol is referred to as a prompt when the computer is waiting for the learner to make a response. Second, a clue that stimulates the learner to give a correct answer is also referred to as a prompt. Prompts of this second type usually appear when the learner has made a mistake or requested help.

Public Domain Software. Software that is not protected by copyright restrictions. It is legal to make copies of public domain software.

RAM. See *Random-Access Memory.*

Random Access Memory (RAM). The temporary portion of the computer's memory. All information in RAM is erased the moment the computer is turned off. Information is loaded into RAM through the keyboard, from a floppy disk or another auxiliary storage device, or through some other input device.

Read-Only Memory (ROM). The permanent portion of the computer's memory. ROM is not erased when the computer is turned off. Part of ROM includes instructions to automatically transfer information into RAM. Information available in ROM is immediately available as soon as the computer is turned on, but ROM cannot be easily modified.

Reinforcement. The strengthening of a behavior by providing pleasant consequences. In computer-assisted instruction, reinforcement usually consists of either positive feedback, some sort of pleasant visual or auditory display, or the opportunity to engage in a pleasant activity (such as an electronic game).

Resolution. The degree of precision or clarity produced in graphic displays. High-resolution graphics are created with a large number of tiny dots on the screen. Low-resolution graphics are created with small squares. Pictures drawn with higher resolution look much more realistic.

Revelatory Use of Computers. The use of the computer to mediate between the learner and a qualitative or quantitative model of a phenomenon under consideration. Examples include simulations, realistic games, and role-playing situations.

ROM. See *Read-Only Memory*.

Shell Program. A computerized drill program into which users can easily insert their own instructional material. For example, a vocabulary shell program might permit the teacher or student to insert words or definitions, and then the computer would provide a drill on those terms.

Simulation. A program that imitates realistic events that would otherwise be impossible or difficult to incorporate into the classroom because the presentation would be expensive, dangerous, time-consuming, unethical, or otherwise impractical.

Software. The instructions and information (program and data) given to the computer to make it perform designated activities. The software is the set of instructions that makes the hardware carry out its appropriate functions.

Spreadsheet. A program that permits the organized entry and tabulation of numerical data in such a way as to provide automatic recalculation of formulas programmed by the user.

Telecommunications. The process of communicating between computers via a modem.

Test Generator. A program that automatically generates tests or quizzes, using items and guidelines entered by the instructor.

Thermistor. A temperature probe used in a microcomputer-based laboratory, which enables the computer to make automatic recordings of temperature in science experiments.

Tutorial. A program that provides instruction on a topic. A tutorial is usually a computerized presentation of branching programmed instruction.

User Friendliness. The ability of a program to accept and respond to input in such a way that the user can easily interpret and make use of the computer's response.

Videodisc. An auxiliary storage device that employs laser technology to present audio and video displays. Videodiscs can be used in combination with a microcomputer for interactive instruction.

Word Processing. A computer program that enables the computer to be used to type and edit documents.

A P P E N D I X B

SOFTWARE SUMMARIES

The following table contains information pertaining to 264 software packages available for science instruction. Each entry begins with a "Title" category followed by the "Publisher." Addresses for publishers can be found in Appendix D.

"Date" indicates the date of publication of the copy of the software we were able to examine. In many cases, these programs will have been revised and updated before you read this table. "Cost" refers to the price of the software at the time we examined it. Prices are likely to change.

"Mgt" refers to the availability of management options, such as record keeping and permitting the teacher to change parameters for individual students. "Mdfy" means that the program can be modified, as by entering vocabulary of unique interest to a specific group of students.

"Low" refers to the lowest grade level at which the software would appropriately be used, usually based on the publisher's information. "High" refers to the highest grade level at which the software would be suitable. "K" is Kindergarten, the numbers 1–12 correspond to grade levels, and "Adult" refers to college or adult learners.

Finally, the "Comments" column briefly describes the topics covered in the program or gives advice about using the software. Many of these programs are described in greater detail elsewhere in this book, and cross references can easily be found in the book's index.

TITLE	PUBLISHER	DATE	COST	MGT	MDFY
Biology					
ACID RAIN	Diversified	1984	$49.95		
ADAPTION IN TWO BIOMES	Prentice-Hall	1986	$69.00	Mgt	
AIDS: THE INVESTIGATION	Marshware	1989	$54.95	Mgt	
AIDS: THE NEW EPIDEMIC	Marshware	1989	$54.95	Mgt	
ANATOMY OF A FISH	Ventura	1986	$49.95	Mgt	
ANIMAL LIFE DATABASES	Sunburst	1987	$59.00	Mgt	
AQUARIUM	Cross	1981	$15.00		Mdfy
BACKYARD BIRDS	MECC	1989	$59.00		Mdfy
BALANCE: PREDATOR–PREY SIMULATION	Diversified	1984	$54.95		
BIOFEEDBACK MICROLAB (MBL)	Queue	1986	$420.00	Mgt	Mdfy
BIOLEARNING SYSTEMS SOFTWARE SERIES	Biolearning	1985	$129.95 each		
BIOLOGY KEYWORD SERIES	Focus Media	1985	$138.95		

LOW	HIGH	COMMENTS
5	12	Tutorial that explores the relationship between power plants and the deterioration of the aquatic environment. Simulation and quiz included.
9	12	Students draw conclusions from graphs and climatograms and examine plant diversity to draw conclusions about adaption of desert/tundra plants.
9	12	A simulation where students unravel events involving AIDS patients.
9	12	A tutorial and quiz that focuses on AIDS. Includes a simulation.
7	Adult	A tutorial and educational game that presents information on the external, internal, and skeletal structures of a bony fish.
4	12	Needs the BANK STREET SCHOOL FILER to access the database. Students can browse, sort, and find; they can print custom reports. Includes a tutorial and reference manual.
2	9	Simulates the life of a community of fish in an aquarium. Several games are included: aquarium fishing, guppy chase, and fish food race.
3	9	Simulates a bird-watching field trip. Students identify birds while on the trip. Students can explore a bird database and access a glossary.
9	12	Program allows manipulation of five variables to investigate predator/prey relationships with tabular and graphic output.
7	Adult	Probes measure skin temperature, heart rate, muscle tension, and electrodermal activity. Students investigate what affects the readings.
7	12	A series of tutorials in biology, including genetic vulnerability, photosynthesis, kidney function, cell respiration, and biochemistry.
7	10	Game of thinking, word power, and spelling to develop an understanding of scientific terms. Includes four programs.

TITLE	PUBLISHER	DATE	COST	MGT	MDFY
BIOLOGY LAB	Cross	1984	$189.95		
BIOLOGY MIND GAMES	Diversified	1985	$39.95		Mdfy
BIOLOGY PROGRAMS SERIES	J & S	1986	$39.00 each	Mgt	Mdfy
BIOLOGY TEST MAKER	J & S	1981	$109.95		Mdfy
BIRDBREED	EduTech	1985	$59.95		Mdfy
BODY DEFENSES	Prentice-Hall	1986	$69.00	Mgt	
BODY ELECTRIC (MBL)	Queue	1989	$450.00	Mgt	
BODY SYSTEMS SERIES	Marshware	1985	$41.95 each	Mgt	Mdfy
BODY TRANSPARENT	DesignWare	1985	$39.95		Mdfy
BOTANICAL GARDENS	Sunburst	1986	$65.00		Mdfy
CARDIOVASCULAR FITNESS LAB (MBL)	Queue	1985	$175.00	Mgt	

LOW	HIGH	COMMENTS
7	10	A package of seven programs covering the dissection of commonly used invertebrates along with the frog and fish. Includes a self-test.
8	Adult	An educational computer board game designed for content review of general biology, for one to four players. Includes a question bank of 140 questions on biologists, discoveries, terms, and organisms.
6	12	A series of drills in biology, including reproduction, excretion, biochemistry, digestion, photosynthesis, respiration, genetics, and locomotion.
9	12	Teacher can make tests or quizzes from a nine-disk, 1,500 question data bank. Can mix questions from different units and randomize questions.
7	Adult	Simulation of the inheritance of color in parakeets. Crosses can be performed between any two birds within a breeding group. Students can investigate dominance, independent assortment, alleles, etc.
9	12	Tutorial that requires students to analyze various body defenses and decide which are categorized as first line.
8	Adult	Students monitor and measure brain waves, electrocardiograms, electrical activity of muscles, and other tests.
2	6	The series includes programs on the endocrine, nervous, digestive, and respiration systems; heart; blood; bones; and muscles. Short tutorials.
5	10	Students move bones and organs to correct locations in the human body or play game of facts and functions. Can add questions to game.
6	12	Simulation of a greenhouse where students can grow plants from seed under various conditions. Students custom design their own seeds/plants.
5	Adult	MBL requiring interface card with light probe, which attaches to earlobe or finger, to measure heartbeat. Also a fitness training section.

TITLE	PUBLISHER	DATE	COST	MGT	MDFY
CATLAB	Conduit		$79.95		
CLASSIFY	Diversified	1983	$54.95		
CLASSIFYING MAMMALS	Prentice-Hall	1986	$69.00	Mgt	
DESIGNER GENES	QED	1988	$49.95		
DICHOTOMOUS KEY TO POND MICROLIFE	EduTech	1989	$59.95		Mdfy
DILUTE	Diversified	1983	$54.95		
DNAGEN	Diversified	1982	$49.95		Mdfy
DNA—THE MASTER MOLECULE: LEVEL I: THE BASICS	EME	1988	$58.95		
DRUGS AND HEARTBEAT	Cross	1987	$29.95		
ELEMENTARY SCIENCE CONCEPTS SERIES	Profiles	1987		Mgt	
ENDANGERED SPECIES DATABASES	Sunburst	1988	$59.00	Mgt	
ENZYME	Diversified	1982	$54.95		

LOW	HIGH	COMMENTS
9	Adult	Students breed domestic cats on the basis of coat color and pattern and observe the transmission of traits through the generations.
7	12	A computerized dichotomous key to the five-kingdom classification system. Students use the program to understand how systems are designed.
9	12	Students construct a classification system, draw conclusions about the various groups, and analyze similarities and differences among mammals.
7	12	Covers Mendelian genetics with monohybrid and dihybrid crosses involving simple and multiple alleles. Includes sex-linked alleles and Punnett squares.
9	12	A computerized key to over seventy-five microlife forms. A database is provided.
9	12	A simulation of the serial dilution technique with known and unknown concentrations of bacteria. Integrated with real lab activities.
9	12	An interactive program that simulates the functioning of the genetic code. Students can generate code and study mutation effects.
9	Adult	Simulates the building of a DNA molecule, the transcription of mRNA, protein synthesis, mutation, and other related phenomena.
6	12	Simulates the actual view through a microscope when various chemicals are put on a daphnia.
4	6	LOOKING INSIDE SERIES includes bones, muscle, and blood. Others include growing seeds. HUMAN BODY SERIES includes the skeletal system.
4	12	Needs the BANK STREET SCHOOL FILER to access the database. Students can browse, sort, and find; they can print custom reports. Includes a tutorial and reference manual.
9	12	Simulation that studies the application of the lock-and-key model of enzyme action. Students compare structures of inhibitors to substrates.

TITLE	PUBLISHER	DATE	COST	MGT	MDFY
EXPERIMENTS IN HUMAN PHYSIOLOGY	HRM	1986	$324.95		Mdfy
EXPLORE A SCIENCE SERIES	Collamore	1987			Mdfy
FIVE SENSES SERIES	Marshware	1985	$45.00 each	Mgt	Mdfy
FLYGEN	Diversified	1983	$49.95		Mdfy
FOOD WEBS: ECOLOGY OF FOOD CHAINS	Diversified	1984	$49.95		
GENETIC ENGINEEERING	Helix	1989	$94.95		
GENETICS PROBLEM SHOP	EME	1988	$49.95		
THE GREAT BIOLOGY KNOWLEDGE RACE	Focus Media	1986	$84.95		
GRIZZLY BEARS	Advanced Ideas	1988	$49.95	Mgt	
HEART	Bergwall	1987	$99.00	Mgt	
HEART ABNORMALITIES AND EKG's	Focus Media	1985	$69.00		
HEART SIMULATOR	Focus Media	1984	$54.95		

LOW	HIGH	COMMENTS
7	12	An MBL package that includes everything needed for ten experiments in human physiology. BASIC programs allow students to design experiments.
2	6	Programs that combine science exploration with reading, writing, and language arts. Students change scenes on the screen and create scenes.
2	6	Series of short tutorials on senses including sight, smell, touch, and sound. Graphic presentations and teacher management.
9	12	Simulates monohybrid or dihybrid crosses in *Drosophila* with twenty-five variations. Sex linkage and chromosomal linkage are included.
6	10	Uses graphics, a four-part tutorial, review game, and quizzes to introduce students to the ecology of food relationships.
10	Adult	An extensive and thorough presentation of cells, genes, proteins, DNA, RNA, and other topics, in a tutorial with graphics.
9	Adult	Presents problems of monohybrid, dihybrid, and trihybrid crosses with human and plant traits, dominant and recessive genes, and genotypes and phenotypes.
9	12	Presents challenging questions drawn from a traditional biology curriculum, for small groups or entire classrooms.
4	12	A simulation where students play the role of a park ranger, a researcher, or a resource developer to resolve conflicts between humans and bears.
7	12	Tutorial on the heart, including its location, chambers, valves, blood vessels, and blood flow. Includes quizzes and management system.
7	12	Has demonstration mode where students compare normal to abnormal heartbeat. A tutorial mode teaches students about heart abnormalities.
6	12	Animated demonstration of blood flow through human heart. Students can control the speed of heartbeat and can use stop action.

TITLE	PUBLISHER	DATE	COST	MGT	MDFY
HEREDITY DOG	HRM	1983	$49.95		
HOWS & WHYS OF MIGRATING MOLECULES	Thorobred	1984			
HUMAN GENETIC DISORDERS	Queue	1988	$49.95		
HUMAN PUMP	Sunburst	1986	$65.00		
INSECT WORLD	Heath	1985	$66.00		
INSECT WORLD	Ventura	1986	$69.95	Mgt	
INVISIBLE BUGS	MECC	1989	$59.00		
LIFE SCIENCE PROGRAMS	J & S	1985	$39.00 each	Mgt	Mdfy
LUNAR GREENHOUSE	MECC	1989	$59.00		
MANRGY	Diversified	1983	$49.95		
MARINE LIFE SERIES	Ventura	1986	$49.95 each	Mgt	
MECC DATAQUEST: NORTH AMERICAN MAMMALS	MECC	1988	$59.00		Mdfy

LOW	HIGH	COMMENTS
7	12	Simulation allows mating of dogs of different coat colors and patterns to produce litters of pups. Single-gene or two-gene systems.
7	Adult	Tutorial on molecules, diffusion, osmosis, equilibrium, and related ideas through direct instruction, hidden words, puzzles, and questions.
9	Adult	Explores Huntington's disease, cystic fibrosis, hemophilia, albinism, genotypes, and phenotypes.
5	12	Tutorial and drill program that teaches about the function and care of the human heart.
4	6	Tutorial game that helps students understand the world of the honeybee or ladybug beetle. Students play the role of the insect.
7	12	Tutorials and practice on insect structure and survival. Includes tutorials, explorations, quizzes, and glossary.
3	9	Students design experiments by selecting beetles and observing their offspring, developing principles of simple inheritance.
4	10	Drill programs in animals, bones/muscles/skin, cell theory, circulation/respiration, control system, digestive system, and ecosystems.
3	6	Futuristic simulation of plant growth. Student can determine the effect of four variables on vegetable germination, growth, and yields.
9	12	An interactive program where students can manipulate the ecological, societal, and population characteristics of a mythical land.
7	12	The series of tutorials include the anatomies of the shark, fish, sea lamprey, and invertebrates. Each includes lessons, "probes," and games.
7	12	Database of facts about ninety-eight North American mammals. Students learn to recognize taxonomic patterns. Use with MECC DATAQUEST COMPOSER.

TITLE	PUBLISHER	DATE	COST	MGT	MDFY
MENDELBUGS	Focus Media	1989	$85.00		
MICRO GARDENER	Educational Activities	1984	$63.00		
MICRO-SCOPE	Cross	1988	$29.95		
MICROBE	Synergistic	1982	$44.95		Mdfy
MONOCROSS, DICROSS	Diversified	1983	$54.95 each		
MOTHS	Diversified	1983	$54.95		
NICHE	Diversified	1982	$49.95		
ODELL LAKE	MECC	1988	$59.00	Mgt	
OPERATION FROG	Scholastic	1984	$79.95		
OSMO	Diversified	1982	$54.95		
OSMOSIS AND DIFFUSION	EME	1985	$52.00		
PLANT	Diversified	1982	$54.95		
PLANT AND ANIMAL CELLS	Ventura	1986	$49.95		

LOW	HIGH	COMMENTS
7	Adult	Students select parent bugs and trace the traits through the generations. Includes Punnett square demonstrations and insect life cycles.
4	9	Students grow geraniums and philodendrons by controlling light, water, temperature, and fertilizer. Has three levels of complexity.
6	10	Program includes short tutorials, matching and multiple-choice tests, and a "Looking Through a Microscope" section with six slides.
7	12	An educational adventure game based on the movie "The Fantastic Voyage." Teams of players must move in a miniature sub to cure a sick person.
7	12	Simulation of various monohybrid and dihybrid crosses.
9	12	Simulation of the evolution of the peppered moth. Population genetics principles are emphasized during the experimentation.
8	12	An interactive program that uses a game format to explore the ecological concept of a niche. Students place organisms in correct niche.
4	6	While role playing various species of fish in a lake, students learn about animal interaction and food chains. Good graphics and animation.
5	12	Simulation of frog dissection with tutorial. Students cut organs from frog, find information about organs/systems, and put organs back.
7	12	Demonstrates a simulated experiment of osmosis in red blood cells.
5	10	Students build a scientific model for osmosis and test factors that affect the movement of substances in and out of cells.
7	12	Simulation of plant growth. Introduces students to the scientific method. Students design, run, and report on simulated growth experiments.
7	12	Tutorial and practice on general structure of plant cells, photosynthesis, animal cells, and mitosis. Generates test questions.

TITLE	PUBLISHER	DATE	COST	MGT	MDFY
PLANT GROWTH SIMULATOR	Focus Media	1988	$98.95 (elementary) $129.95 (advanced)		
PLANT: NATURE'S FOOD FACTORY	Ventura	1985	$69.95	Mgt	
POLLUTE	Diversified	1984	$54.95		
POPGEN	Diversified	1983	$49.95		
POPGRO	Diversified	1982	$49.95		
PRENTICE-HALL BIOLOGY COURSEWARE	Prentice-Hall	1988	$69.00 each	Mgt	
PROJECT CLASSIFY SERIES: PLANTS & MAMMALS	National Geographic	1989	$139.95 each		
PROJECT ZOO	National Geographic	1987	$149.95		
PROTOZOA	Ventura	1988	$39.95	Mgt	
SCIENCE TOOL KIT: MODULE 3 BODY LAB (MBL)	Broderbund	1987	$49.95	Mgt	Mdfy
SENSES: THE PHYSIOLOGY OF HUMAN SENSE	Ventura	1988	$39.95	Mgt	

LOW	HIGH	COMMENTS
4	8	A microworld in which students design plant-growth experiments with variables such as wavelengths of light, amount of moisture, and amount of CO_2.
7	12	
7	12	Tutorials and experiments in plant structure, growth, and genetics. Includes quizzes and a glossary.
9	12	Simulation of effects of pollutants on streams, lakes, and ponds under various conditions. Tabular and graphical data are produced.
9	12	Interactive program centered around the Hardy–Weinburg law and the complex concept of genes in populations.
9	12	A simulation that allows students to change variables that deal with population growth and to study progressive refinement of scientific models.
9	12	Series on the circulatory, reproductive, and respiratory systems. Each tutorial consists of a preview, instruction, and evaluation section.
4	8	Students play the roles of the botanist's apprentice and the field zoologist to meet over seventy-two plant challenges and thirty-six animal challenges.
3	9	Multimedia kit that teaches about zoo animals, skills related to making and interpreting graphs and tables, and zoo design.
7	12	Tutorials and practice on the anatomical structures and functions of microorganisms. Includes data-retrieval utility, quiz machine, and identification game.
3	12	Measures heart rate, lung capacity, and reaction time. Includes manual and cardboard spirometer. Order the school edition.
7	12	Tutorials and practice on the eye, tongue, ear, skin, and nose. Includes an identification game, data-retrieval utility, and quiz matching.

TITLE	PUBLISHER	DATE	COST	MGT	MDFY
SEXUALLY TRANSMITTED DISEASES	Marshware	1988	$88.95		
SURVIVAL OF THE FITTEST	EME	1987	$54.95	Mgt	
VISIFROG	Ventura	1988	$59.95	Mgt	
WEEDS TO TREES	MECC	1989	$59.00		
WHALES	Advanced Ideas		$49.95		
WHO AM I? SERIES	Focus Media	1986	$45.00		
WORM	Ventura	1988	$39.95	Mgt	

Chemistry

TITLE	PUBLISHER	DATE	COST	MGT	MDFY
ALL ABOUT MATTER	Ventura	1987	$49.95		
ATOMIC FORMULAS AND MOLECULAR WEIGHTS	COMPress	1983			
BOYLE'S AND CHARLES' LAWS	Prentice-Hall	1988		Mgt	
CHEM DEMO	CDL	1984			
CHEM LAB	Simon & Schuster	1985	$39.95		

LOW	HIGH	COMMENTS
7	Adult	Presents and tests students on facts of sexually transmitted diseases, including AIDS.
8	Adult	Students capture as many prey as possible while directing the evolution of the predator. The computer guides the evolution of the prey.
6	Adult	Graphical display of frog anatomy, with database of structure and function. Includes a game that challenges students to match structures and functions.
3	9	A realistic simulation of plant succession. Students select and place nine different kinds of plants in a field and observe changes.
4	12	An Audubon Wildlife Adventure; requires 128K of memory.
3	6	An identification game of clues and pictures, where the computer pretends to be an organism. Packages: DESERT ANIMALS, BIRD, TREES & FLOWERS.
7	12	Tutorials and practice on the worm anatomy. Includes identification game, data-retrieval utility, and quiz machine.
4	7	Information presentation with graphics and drills through games and quizzes. Color monitor needed.
9	Adult	Tutorial with drill on molecules, including atomic and molecular weights and gram molecular weights.
9	12	Boyle's, Charles', Gay-Lussac's, combined gas law, and others. Shows how laws were developed; includes tutorial, drill, and conceptual understandings.
9	Adult	Simulations of acid/base titration, atomic absorption spectra, and infrared spectra.
6	12	Using authentic principles, students conduct experiments with thousands of possible combinations of chemicals to produce such compounds as gasoline additives and synthetic diamonds.

TITLE	PUBLISHER	DATE	COST	MGT	MDFY
CHEM LAB SIMULATION 1	High Technology	1979	$100.00		
CHEM LAB SIMULATION 2	High Technology	1979	$100.00		
CHEM LAB SIMULATION 3	High Technology	1981	$100.00		
CHEM LAB SIMULATION 4	High Technology	1981	$100.00		
CHEMAID	Ventura	1986	$49.95	Mgt	
CHEMICAL NOMENCLATURE/ BALANCING EQUATIONS	Bergwall	1984	$59.00		
CHEMISTRY ACCORDING TO ROF	Richard O. Fee	1984	$29.95/ disk	Mgt	Mdfy
CHEMISTRY: THE PERIODIC TABLE	MECC	1988	$58.95	Mgt	
CHEMPAC (MBL)	E & L	1986	$1045.00	Mgt	Mdfy
ELECTRON PROPERTIES	J & S		$39.95		
ELEMENTS	COMPress	1983			
EXPERIMENTS IN CHEMISTRY (MBL)	CDL	1984	$375.00	Mgt	Mdfy

LOW	HIGH	COMMENTS
9	Adult	Simulations of acid/base titrations, equilibrium constant, and strong acid/base. Students titrate on screen and calculate off screen.
9	Adult	Ideal gas law simulation. Paddles needed. Shows constant and random motion with elastic collisions.
9	Adult	Calorimetry. Heat of capacity of calorimeter and heat of neutralization. Hess's law. Simulates heat of hot and cold water, HCl + NaOH pellets, etc.
9	Adult	Thermodynamics. Capillary tube experiments illustrate heat of vaporization and thermodynamics of an equilibrium reaction. Includes enthalpy, entropy, and free energy.
6	12	Introduction to the periodic table, identification, spelling, atomic weight, number, electrons, families, etc. Includes a database.
9	Adult	Tutorial and drill and practice on naming, writing, and balancing equations.
9	Adult	Sixty-eight programs on eighteen disks covering all of general chemistry in tutorial, drill-and-practice, and simulation formats. Good questions.
9	Adult	Students acquire information about elements through "Property Search" and "Element Inquiry."
9	Adult	A complete laboratory program that uses the computer to sense, measure, and record data. Includes texts and equipment.
9	Adult	Programs include the basic properties of electrons, energy levels, and electron configuration.
9	Adult	Tutorial and drill on the periodic table, names of the elements, isotopes, properties of elements, and some mystery ones to identify.
9	Adult	Temperature probes and interface boxes to measure, record, and analyze data from chemistry experiments. Requires a pH electrode.

TITLE	PUBLISHER	DATE	COST	MGT	MDFY
EZ-CHEM GAS LAW SIMULATION	CDL	1985	$99.95		
FAMILIES OF ATOMS	J & S	1982		Mgt	Mdfy
GAS LAWS AND THE MOLE	Prentice-Hall	1988			
GENERAL CHEMISTRY DEMO	COMPress	1984			
INORGANIC NOMENCLATURE	COMPress	1984			
INTERFACING COLORIMETRY PROGRAM (MBL)	Kemtec	1985	$128.60 (green) $233.80 (green/red)	Mgt	
INTRODUCTION TO GENERAL CHEMISTRY	COMPress	1990	$70.00 (per disk)		
IT'S A GAS	Diversified	1985	$64.95		
MOLECULES AND ATOMS	Thorobred	1983			
OIL DROP	EduTech	1988		Mgt	
ORGANIC QUALITATIVE ANALYSIS	COMPress	1984			

LOW	HIGH	COMMENTS
9	12	Simulations of ideal gases, diffusion, mixing, and $PV = nRT$ machine. Can change number of moles, volume, temperature, and pressure. Very good for class discussion.
9	12	This is a question-and-answer program with graphs. The teacher can change questions and adjust the grade level.
9	Adult	Programmed learning with animated visuals. Expansion and diffusion of gases, Dalton's law, combining volumes, mole, molar volume, and ideal and real gases.
9	Adult	Tutorial and drill on the periodic table, atomic structure, inorganic nomenclature, balancing equations, ideal gases, pH, and acid/base titrations.
9	Adult	A tutorial with drill that covers binary salts, variable oxidation states, nonmetals, acids and bases, ternary salts.
9	Adult	Game port connector to Kemtec colorimeter functioning as a spectrophotometer at 565 nm. Measures common absorbancies.
9	Adult	Tutorial series of ten disks designed to cover general chemistry.
9	12	Ideal gas law demonstration, gas law quiz and problems, refrigerator simulation, and CO_2 fire extinguisher. Good visual explanation of gas laws.
7	Adult	Tutorial on the Bohr model of atom, nature of molecules, atomic mass and number, charges, and formation of chemical compounds.
9	Adult	Paddles are used to control voltage in simulated Millikan oil drop experiment. Two levels of difficulty. Students record and pool data.
9	Adult	Allows students to perform tests on knowns to see what happens and then gives unknowns to identify.

TITLE	PUBLISHER	DATE	COST	MGT	MDFY
PERICHART	Intellimation	1986	$26.95		
pH PLOT (MBL)	CDL	1983	$74.95	Mgt	Mdfy
PROJECT SERAPHIM	Project SERAPHIM	Various	Various	Various	Various
Earth Science					
ALL ABOUT THE SOLAR SYSTEM	Ventura	1987	$49.95		
ASTRONOMY DATABASES	Sunburst	1987	$59.00	Mgt	
ASTRONOMY: STARS FOR ALL SEASONS	Educational Activities	1983	$49.95		
CELESTIAL BASIC	American Only	1984	$49.95 (2 disks)		
CLIMATE AND WEATHER DATABASES	Sunburst	1988	$59.00	Mgt	
CONTINENTAL DRIFT	Prentice-Hall	1986			
DINOSAUR DISCOVERY	EME	1987		Mgt	
THE EARTH AND MOON SIMULATOR	Focus Media		$98.95		
ENERGY SEARCH	Macmillan McGraw-Hill	1982	$235.00	Mgt	

LOW	HIGH	COMMENTS
10	Adult	Database of atoms and elements that can be searched through interaction with a graphic periodic chart. Requires a Macintosh computer.
9	Adult	pH-probe interfaced. Measures or produces and displays both simulated and experimental pH data. Data can also be entered from keyboard.
9	Adult	Includes all chemistry subject areas, with software, databases, and interfacing programs.
4	7	Information presentation with graphics and drills through games and quizzes. Requires a color monitor.
4	12	Databases contain pre- and post-1800 events, planets, logbooks, glossary, constellations. Use with BANK STREET SCHOOL FILER.
4	12	Tutorial on appearance of the sky and graphical views of the nighttime sky as seen from a backyard.
9	Adult	Moon, planets, phases, location, and eclipses. Also discusses calendars and dates.
4	12	Needs the BANK STREET SCHOOL FILER to access the database. Students can browse, sort, and find; they can print custom reports. Includes a tutorial and reference manual.
6	12	Map of ocean floor; not hands-on. Includes a test and crossword puzzles.
4	7	A language adventure game with pictures. Students attempt to find and hatch a dinosaur egg. They must keep notes and perform experiments.
5	12	A simulation of the paths of the Earth and moon, phases of the moon, rotation and revolution, sideral and synodic months, eclipses, and more.
5	12	Students are directors of an energy factory, which they manage. Animal, water, wood, coal, oil, nuclear, and solar energy sources. Includes student packet.

TITLE	PUBLISHER	DATE	COST	MGT	MDFY
GEOLOGICAL HISTORY	Sunset	1987	$65.00	Mgt	
GEOLOGISTS AT WORK	Sunburst	1987	$65.00	Mgt	
GEOLOGY IN ACTION	CDL	1987	$49.95	Mgt	
GEOLOGY SEARCH	Macmillan McGraw-Hill	1982	$235.00	Mgt	
GEOSTRUCTURES	Intellimation	1986	$12.95	Mgt	
GEOWORLD	Tom Snyder Productions	1986	$79.95	Mgt	Mdfy
HALLEY'S COMET	American Only	1985	$49.95		
IGNEOUS ROCKS	Ward's Natural Science	1984	$67.00	Mgt	
INTERPLANETARY TRAVEL	Prentice-Hall	1986		Mgt	
METAMORPHIC ROCKS	Ward's Natural Science	1984	$67.00	Mgt	
MINERAL DATABASE	Sunburst	1989	$59.00	Mgt	
MINERAL IDENTIFICATION COMPUTER PROGRAM	Scott Resources	1986	$99.00	Mgt	Mdfy

LOW	HIGH	COMMENTS
7	12	Students investigate deposition, folding, faulting, tilting, intrusions, and erosion; analyze and interpret cross sections. Program is inquiry-oriented.
7	Adult	Students work with Earth processes to create and interpret cross sections and then three-dimensional models and maps from them and analyze core samples.
6	Adult	A tutorial and experimentation about geological processes and cross sections allowing interpretation and creation of cross sections.
7	12	Student groups search for oil on an island continent using various geological tests and trial drilling. Includes teacher/student manuals.
9	Adult	Students tilt, fold, fault three-dimensional block diagrams and observe structures to learn to interpret Earth history. Requires a Macintosh computer.
7	12	Mineral resource distribution based on a database of fifteen natural resources. Can be transferred to APPLEWORKS. Game and research versions.
9	Adult	Amateur astronomers look at Comet Halley in history, when it has appeared, how to find its path, how its returns were confirmed, and more.
7	12	Identification of igneous rocks and processes. Includes a set of nine igneous rocks and a guide. Tutorial has good graphics but requires a lot of reading.
6	12	Tutorial instruction; stops and asks questions on planetary data, but it is out-of-date. Provides good thought questions.
7	12	Identification of metamorphic rocks and processes. Includes a set of nine metamorphic rocks and a guide. Tutorial has good graphics but requires a lot of reading.
6	12	Needs the BANK STREET SCHOOL FILER to access the database. Students can browse, sort, and find; they can print custom reports. Includes a tutorial and reference manual.
7	12	Identification of rocks and minerals using testable properties. Includes four sets of rocks and a guide. Students perform tests and use a branching program to identify.

TITLE	PUBLISHER	DATE	COST	MGT	MDFY
PLANETARY CONSTRUCTION SET	Sunburst	1987	$59.00		
PLATE TECTONICS	Educational Images	1986	$59.95	Mgt	
ROCK CYCLE	Ward's Natural Science	1984	$67.00	Mgt	
ROCKY: THE MINERAL IDENTIFICATION PROGRAM	Michigan State University	1986	$50.00		
SCIENCE 4: SPACE—UNDERSTANDING OUR SOLAR SYSTEM	Decision Development	1987			
SCIENCE TOOL KIT: MODULE 2 EARTHQUAKE LAB (MBL)	Broderbund	1986	$49.95	Mgt	Mdfy
SEDIMENTARY ROCKS	Ward's Natural Science	1982	$67.00		
SIMULATIONS IN EARTH SCIENCE	Data Command	1988	$39.95 each	Mgt	
SKY LAB	MECC	1985	$59.00		
SOLARISM	Interstel	1987			Mdfy
SPACE DATABASES	Sunburst	1987	$59.00	Mgt	

LOW	HIGH	COMMENTS
6	12	Students design a planet to fit characteristics of aliens.
7	12	This tutorial, on two 800K disks, demonstrates concepts and shows involvement of plates. Historic, current, and future positions are evident. Includes hot spots, isochrons, and earthquake locations.
4	12	Tutorial with lots of reading on the origin of rocks and minerals.
7	Adult	Identification of mineral hand specimens by entering observable properties; computer eliminates those lacking property from list.
4	7	Covers mapping, energy in space, small objects, planets, solar system, galaxies, and deep space.
4	12	Earthquake location and measurement. Seismoscope with manual. Order the school edition.
7	12	Identification of sedimentary rocks and processes. Includes a set of nine sedimentary rocks and a guide. Tutorial with graphics, a lot of reading, and very directed.
3	9	Visual tutorials with questions and branching to help with earthquakes, volcanoes, clouds, and weather instruments (four disks).
7	9	Students observe the stars and learn about the relationship between the movement of the Earth, sun, stars, and planets. (Received the Electronic Learning Award.)
9	Adult	Planetary motion, Comet Halley, and location of stars, planets, and others. Shows right ascension, declination, azimuth, and altitude.
4	12	Needs the BANK STREET SCHOOL FILER to access the database. Students can browse, sort, and find; they can print custom reports. Includes a tutorial and reference manual.

TITLE	PUBLISHER	DATE	COST	MGT	MDFY
STAR SEARCH	Earthware	1983	$49.00		
TELLSTAR: COMET HALLEY EDITION, NEW VERSION	Spectrum Holobyte	1984	$14.95		Mdfy
TIME AND SEASONS	Rand McNally	1983	$111.00		
TIMELINER	Tom Snyder Productions	1986	$59.95		Mdfy
VOLCANOES	Earthware	1988	$49.50		
VOYAGER VERSION 1.2	Carina	1988	$119.95		Mdfy
WEATHER FRONTS	Diversified Educational Enterprises	1984	$49.95		
WEATHER OR NOT	Rand McNally	1984			
WEATHER PRO	Petrocci	1987	$34.95	Mgt	
WEATHER STAT	Petrocci	1987	$34.95		
WEATHER TRACK	Petrocci	1987	$34.95	Mgt	

LOW	HIGH	COMMENTS
4	12	Students simulate a mission in which they test for possible life forms on other worlds, land, and make contact. Mockingboard is helpful.
7	Adult	Location and identification of planets, stars, constellations, and deep-sky objects. Shows ancient to modern skies.
7	9	Tutorial and practice on meridians, longitude lines, timelines, celestial meridians, time zones, rotation, seasons, and length of day.
K	12	Creation, display, printing, and interpretation of timelines. Five different data disks.
7	Adult	Simulation, prediction, and consequences of volcanic eruptions.
9	Adult	Interactive desktop planetarium, which accurately shows stars, planets, and other objects from Earth, moon, and various other perspectives.
4	12	Tutorial and practice with weather symbols, weather maps, and prediction of the weather. Superior hardcopy manual.
4	9	Questions on what causes and affects the weather, with graphics. A game about the Iditerod race is included as a separate disk.
9	Adult	Area-specific forecaster. Forecasts highs, lows, rainfall, and others. Data entry required. Interactive with WEATHER TRACK.
9	Adult	A statistical program with weather information for an entire year. Works with WEATHER TRACK to calculate averages and other stats. Requires an IBM computer.
9	Adult	Users input daily weather and output display as monthly report. Needs IBM system programs. Works with WEATHER PRO and WEATHER STAT.

TITLE	PUBLISHER	DATE	COST	MGT	MDFY
Elementary Science					
ANIMAL SCIENTIST	Scholastic		$39.00		
ANIMAL TRACKERS	Sunburst	1989	$65.00		
ANT FARM	Sunburst	1987	$65.00		
THE DESERT	Collamore	1987	$75.00		Mdfy
ELECTRICAL CELLS	Macmillan		$39.00		
EXPLORING TIDEPOOLS	Disney Educational Enterprises		$75.00		
GROWING SERIES: SEEDS	Profiles	1987	$59.00	Mgt	
MYSTERY OBJECTS	MECC	1988	$59.00		
PLANT DOCTOR	Scholastic		$39.00		
WHAT WILL HAPPEN?	Macmillan		$39.00		
General Science and Multiple Topics					
BAFFLES	Conduit	1983			

LOW	HIGH	COMMENTS
3	6	As members of a team studying endangered species, students learn to use a database. The program teaches students to identify and classify animals.
4	8	Students explore habitats and try to identify animals in each. Students must look for evidence of the animal's presence.
3	8	Ants travel in mazes or workstations, and students must deduce the shape of the maze.
2	5	Students gather information about desert plants and animals. Students can create their own screens and print information files.
4	7	Students discover how electrical cells work, by working in a simulated lab and converting electrical energy to other forms of energy.
4	8	Through investigating different levels of a tidepool, students develop knowledge of life cycles and predator/prey relationships.
1	5	A tutorial on language skills and science content objectives. Students use interactive learning with questions and feedback.
2	4	Students develop inquiry process skills by testing hidden objects for physical properties such as texture, size, smell, weight, shape, and color.
2	6	Students are Dr. Green's assistants as they "cure" sick plants. Students must develop and test hypotheses in an experimental approach.
1	4	Students use observations to make predictions about simulated events.
9	12	Interactive game requiring deductive reasoning. Students predict location of hidden "baffles" as a result of light reflection in black box.

TITLE	PUBLISHER	DATE	COST	MGT	MDFY
CHALLENGE SERIES	Island	1985			Mdfy
DATA ANALYSIS	EduTech	1985		Mgt	
DATATECH MENTOR MASTER	Data Tech	1985		Mgt	Mdfy
DISCOVERY	Milliken	1985	$150.00	Mgt	
DISCOVERY LAB	MECC	1984			
THE GAME SHOW	Advanced Ideas	1983			Mdfy
GRAPHICAL ANALYSIS III	Vernier	1988	$29.95		
HOW TO BUILD A BETTER MOUSETRAP (MBL)	Vernier	1986	$24.95		Mdfy
KIDNET (KIDS NETWORK)	National Geographic	1989	$472.50 each	Mgt	Mdfy
LAB STATISTICS	High Technology	1982	$75.00		
MACSCOPE (MBL)	Thornton	1989	$1750.00 (without probes)	Mgt	Mdfy
MATHEMATICS FOR SCIENCE	Merlan	1988	$216.95		

LOW	HIGH	COMMENTS
3	Adult	On-screen drill for two people (or teams) to compete to answer questions in any area. Teacher can make own databases.
7	12	Requires printer with graphics interface. Analysis of experimental data; e.g., mass, density, circumference, population growth, and lenses.
K	Adult	Teacher creates units with pictures from files. Is menu-driven. CAI authoring program. Includes meteorology, space, oceanography, cells, and anatomy.
4	12	Presents ten interesting problems for students to solve. Each requires student to plan, predict, experiment, and analyze results.
5	12	Students perform simulated experiment on organisms returned from outer space. Students control variables and test hypotheses.
6	14	A game show format for students in topics that include science, math, language arts, and social studies. Has an authoring program to create individual topics.
9	Adult	Graphs experimental data. Includes manual.
9	Adult	Project manual with disk that shows fourteen interfacing projects, which you must buy and build yourself.
4	6	An innovative telecommunications-based science curriculum, made up of six-week units, incorporating hands-on experiments and computers.
9	Adult	Instructive program on which statistics to use; allows input of data and various functions to be fit to the data. Uses significant figures.
9	Adult	This is an eight-channel A to D interface that features data acquisition and fast sampling speed, with data file storage for a cricket graph.
7	12	A four-series set of lessons and drills on mathematics concepts for science students. The series are available separately for $74.95.

TITLE	PUBLISHER	DATE	COST	MGT	MDFY
THE MBL PROJECT: HEAT AND TEMPERATURE	Queue	1987	$355.00 (includes interface card)	Mgt	Mdfy
ORBITER	Spectrum Holobyte	1986	$49.95	Mgt	
PERSONAL SCIENCE LABORATORY (MBL)	IBM	1989	$357.00	Mgt	Mdfy
PLAYING WITH SCIENCE: TEMPERATURE (MBL)	Sunburst	1988	$99.00	Mgt	Mdfy
SCIENCE BASEBALL GAMES: EARTH SCIENCE	J & S	1988	$39.50		Mdfy
SCIENCE SKILL DEVELOPMENT	EduTech	1988	$35.00		
SCIENCE TOOLKIT: MASTER MODULE (MBL)	Broderbund	1986	$99.95	Mgt	Mdfy
SCI-LAB (MBL)	Sargent-Welch	1984	$1695.00	Mgt	
SECOND VOYAGE OF THE MIMI	Sunburst	1989	$1050.00		
THE SPI SYSTEM (MBL)	Thornton	1987	$995 (plus probe)	Mgt	Mdfy
TEMPERATURE EXPERIMENTS (MBL)	Hartley	1985	$69.95	Mgt	Mdfy

LOW	HIGH	COMMENTS
6	Adult	Using thermistor, the program measures, records, and analyzes data. Designed to help students explore and challenge intuitive ideas of temperature and heat.
7	Adult	Space shuttle simulation with a manual. Students fly the space shuttle and complete voyages. Requires a Macintosh computer.
K	Adult	Box interfaces to IBM with probes for temperature, light, pH, distance, and others.
K	Adult	A study of temperature utilizing thermistors plugged into game port. Includes three thermistors, cables, and interface.
6	12	Drills in earth science, physics, chemistry, physical science where student gets hits and scores runs in a baseball game format.
5	10	A skills practice program on reading scales, significant figures, scientific notation, multiplication, and division.
4	12	Study of time, temperature, light intensity. Includes an MBL box, thermistor, and manual with experiments. Order the school edition.
9	Adult	A set of eighteen biology, chemistry, and physics interfacing laboratories with excellent graphics. Requires an I/O port pin and an A/D card and box.
4	9	Students explore astronomy, archeology, history, numbers, and calculations. The package includes videotapes, learning modules, software, and teacher's guide.
7	Adult	Interface systems designed for many different experiments. Includes probes and additional hardware. Used with IBM, Macintosh, and Apple II family.
4	12	Simple MBL interface with two temperature probes. Includes a tutorial on learning to read thermometers. Graphs and experimental information will print.

TITLE	PUBLISHER	DATE	COST	MGT	MDFY
TEMPERATURE PLOTTER III (MBL)	Vernier	1988	$39.95		
VISUAL ILLUSIONS	CDL	1987	$68.75	Mgt	Mdfy
VOYAGE OF THE MIMI	Sunburst	1986	$1300.00		

Physics

TITLE	PUBLISHER	DATE	COST	MGT	MDFY
AEROTREK	Estes	1988			Mdfy
AIR TRACK SIMULATOR	CDL	1983	$59.95		
ALL ABOUT LIGHT AND SOUND	Ventura	1987	$49.95		
ALL ABOUT SIMPLE MACHINES	Ventura	1987	$49.95		
ALTERNATING CURRENT	Bergwall	1987	$344.00	Mgt	
ASTROCAD: PERFORMANCE ANALYSIS OF ROCKETS	Estes	1987	$19.95		Mdfy
BASIC ELECTRICITY—DC	Bergwall	1987	$344.00	Mgt	
CHARGED PARTICLE WORKSHOP	High Technology	1982	$75.00		

LOW	HIGH	COMMENTS
9	12	Temperature interfacing with thermistors from a two-probe temperature system. Build or buy interface and probes.
6	Adult	A unit on scientific problem solving. Students gather information on T-illusion, Muller–Lyer, parallelogram, and others and graph, analyze, and interpret.
4	9	Students explore maps and navigation and study whales and ecosystems in the Atlantic Ocean. The package includes a videotape, print, software, learning modules, and guides.
9	Adult	Students enter model parameters, and program predicts stages, altitude, and others. Also predicts altitudes if launched on the moon.
10	Adult	Tutorial on velocity, momentum, energy, and simulations of possible collisions of two objects on horizontal, frictionless surface (air track).
4	8	Information presentation with graphics and drills through games and quizzes. Requires a color monitor.
4	8	Information presentation with graphics and drills through games and quizzes. Requires a color monitor.
10	Adult	Programmed learning with lots of reading. Thirteen disks cover current, voltage, magnetism, inductors, transformers, capacitors, etc.
8	Adult	Calculations analyzing model-rocket design and performance. Technical.
10	Adult	Programmed learning with lots of reading on electron theory, circuits, symbols, diagrams, meters, Ohm's law, series and parallel circuits, and resistors.
10	Adult	Three programs simulating motion of charged particle under the influence of electric and magnetic forces.

TITLE	PUBLISHER	DATE	COST	MGT	MDFY
CHARGED PARTICLES II	Vernier	1983	$24.95		
CIRCUIT LAB	Mark Davids	1984	$30.00		Mdfy
FLIGHT: AERODYNAMICS OF A MODEL ROCKET	Estes	1986	$44.95		
FREQUENCY METER (MBL)	Vernier	1984	$39.95	Mgt	
GENERAL PHYSICS	COMPress	1985	$495.00 each		
GENERAL PHYSICS	Cross	1986	$249.95		
HARMONIC MOTION WORKSHOP	High Technology	1982	$75.00		
IN SEARCH OF SPACE: MODEL ROCKETS	Estes	1986	$24.95		
INTERACTIVE OPTICS	EduTech	1989	$74.95		
KINEMATICS II	Vernier	1983	$24.95		
LAWS OF MOTION	EME	1982	$118.00		
MASS SPRING	CDL	1984			

LOW	HIGH	COMMENTS
9	Adult	Users "experiment" with charged particles in magnetic and electric fields. These are graphical representations. Includes manual.
8	Adult	Inquiry-based look at DC circuits by building and testing own graphic series, parallel, and combination circuits.
7	Adult	Forces on flying objects, satellites, space shuttle, etc. Program explains why objects are able to fly.
8	12	Game port interface allowing input of sound, analysis of sound frequencies, and demonstrations of physics of music. Users can build or buy.
10	Adult	Tutorials on mechanics, heat, waves, sound, light, and more.
10	Adult	Includes twelve volumes on vectors and graphing, statics, motion, circular motion, conservation laws, thermodynamics, electricity, magnetism, and more.
10	Adult	Visually presents simple and damped harmonic motion with high-resolution graphics. Students alter phase, amplitude, and damping factor and see effects.
7	Adult	Flight profiles, parts, how to construct, types of engines, etc.
7	12	Geometric optics principles using simulations and interactive demonstrations. Covers lenses, mirrors, Fermat's principle, and Snell's law.
9	Adult	Student controls motion of a truck according to an assignment. After student enters values, truck moves, and motion is plotted and analyzed.
6	12	Inquiry simulation about motion of blocks along surface and on inclined plane. Simulated motion in other "worlds" with different laws.
9	Adult	Basic math and science concepts applied to real-world systems—shocks on cars, hinged screen door, and beam balance.

TITLE	PUBLISHER	DATE	COST	MGT	MDFY
THE MBL PROJECT: MOTION	Queue	1987	$199.95 (includes interface card)	Mgt	Mdfy
THE MBL PROJECT: SOUND	Queue	1987	$199.95 (includes interface card)	Mgt	Mdfy
MILLIKAN OIL DROP EXPERIMENT	Vernier	1988	$29.95		
MINER'S CAVE	MECC	1988	$58.95	Mgt	
NEWTON'S FIRST LAW	Prentice-Hall	1986		Mgt	
ORBIT II	Vernier	1983	$24.95		
PHYSICAL SCIENCE SERIES: HEAT & LIGHT	Educational Activities	1986		Mgt	Mdfy
PHYSICS LAB: LIGHT (MBL)	Cross	1983	$60.00		
PHYSICS OF MODEL ROCKETRY	Estes	1987	$24.95		
PLAYING WITH SCIENCE: MOTION	Sunburst	1989	$165.00		Mdfy
PRECISION TIMER II (MBL)	Vernier	1985	$39.95		
PROJECTILE MOTION WORKSHOP	High Technology	1982	$75.00		

LOW	HIGH	COMMENTS
7	Adult	A study of distance, velocity, and acceleration, using a motion detector interfaced to the computer.
7	Adult	Microphone interfaced to collect, record, and analyze data to explore the qualitative aspects of amplitude, frequency, wavelength, and wave shapes.
10	Adult	Allows simulation of Millikan's classic experiment. Microscopic view of area between charged plates. Controlled with keystrokes.
5	12	Students manipulate levers, pulleys, ramps, wheels, and axles to move objects. Intuitive study of efficiency of machines in inquiry mode.
7	12	Inertia, forces, and Newton's first law.
9	Adult	Orbital motion, velocity, and force. Includes manual.
4	12	Study of heat and light through tutorial, animation, and simulation.
9	Adult	Light intensity, acceleration, pendulum. Includes four phototransistors, cable, DIP plug, and manual.
9	Adult	Parts of model rockets, forces, laws of motion, acceleration, and gravity.
4	8	Teaching and learning tool as students discover the relationship between movement and lines on position graph. The teacher can control the learning level.
9	Adult	A study of timing, speed, and acceleration using photogates interfaced to game port. Includes manual. Build or buy interface and probes.
10	Adult	Four programs illustrating motion under influence of gravity: dropped vertically, fired upward, fired downward, and component motion.

TITLE	PUBLISHER	DATE	COST	MGT	MDFY
PROJECTILES II	Vernier	1983	$24.95		Mdfy
RAY TRACER	Vernier	1982	$24.95		
SATELLITE ORBITS	Conduit	1983			Mdfy
SCIENCE TOOL KIT: MODULE I: SPEED AND MOTION (MBL)	Broderbund	1986	$49.95	Mgt	Mdfy
SIR ISAAC NEWTON'S GAMES	Sunburst	1985	$65.00		
STANDING WAVE WORKSHOP	High Technology	1982	$75.00		
TUTORIAL ON ACCELERATION	COMPress	1987			
TUTORIAL ON MOTION	COMPress	1988			
VECTOR ADDITION III	Vernier	1988	$29.95		
VOLTAGE PLOTTER III (MBL)	Vernier	1988	$39.95		
WAVE ADDITION III	Vernier	1983	$29.95		
WAVES	J & S	1987	$27.95	Mgt	

Problem Solving

| CAR BUILDER | Weekly Reader | 1985 | $39.95 | | |

LOW	HIGH	COMMENTS
9	Adult	Nine challenges to experiment with projectile motion. Projectile motion is graphically displayed as height, range, and elapsed time.
9	Adult	Students draw ray diagrams to illustrate reflection; includes dispersion, Snell's law, mirrors, lenses, images, and aberration.
10	Adult	A Chelsea science simulation following the Nuffield physics course. Students vary velocity and height to try to produce circular orbit.
4	12	Shows time, speed, and acceleration. Includes car, balloons, light probe, and manual with experiments. Order the school edition.
4	9	Intuitive study of motion around a track with ice, sand, or grass. The track can be on earth, near the sun, or in outer space.
10	Adult	Simulates motion of vibrating string. Shows first ten harmonics individually or superimposed, and students pluck at various positions and analyze.
9	Adult	Tutorials with graphics and tests on one-dimensional motion.
9	Adult	Tutorials with graphics and tests on position and velocity.
9	12	Graphically adds or subtracts vectors. Draws resultants.
9	12	Monitors and analyzes voltage signal from voltage input unit or from advanced interfacing board. Build or buy interface and probes.
9	Adult	Graphically demonstrates superposition of waves. Has simulation and input modes. Demonstrates wavelength, frequency, wave speed, interference, etc.
9	Adult	The program has graphics and experimental setups.
4	8	Students design, construct, refine, and test cars they build on the screen.

TITLE	PUBLISHER	DATE	COST	MGT	MDFY
DISCOVER: A SCIENCE EXPERIMENT	Sunburst	1986	$65.00	Mgt	
DISCOVERY! EXPERIENCES WITH SCIENTIFIC REASONING	Milliken	1982	$150.00	Mgt	
THE FACTORY	Sunburst	1990	$65.00		
GEOMETRIC SUPPOSER	Sunburst	1985	$99.00		
GERTRUDE'S PUZZLES	Learning Company	1985	$59.95		Mdfy
GERTRUDE'S SECRETS	Learning Company	1985	$59.95		
THE INCREDIBLE LAB	Sunburst	1986	$65.00		
KING'S RULE	Sunburst	1985	$65.00		
THE POND	Sunburst	1984	$65.00		
SCIENTIFIC REASONING SERIES	IBM	1986	$39.00–$54.00 each (5 programs)		
SECRETS OF SCIENCE ISLAND	Grolier	1985	$59.95		
WOOD CAR RALLY	MECC	1988	$59.00		

LOW	HIGH	COMMENTS
6	12	Students investigate the behavior of small creatures in a laboratory environment, trying to discover what they like to eat, how they behave, and the like.
4	12	Students increase problem-solving skills by planning, predicting, experimenting, testing, and analyzing results. The program contains ten multilevel problems.
4	8	Students create geometric puzzles on a simulated conveyor belt. They must use inductive thinking, visual discrimination, and spatial perception.
6	12	Students draw, measure, and repeat constructions to collect visual and numerical data as they discover geometry.
3	7	Students try to solve more advanced problems similar to those in GERTRUDE'S SECRETS.
K	4	Gertrude, a goose, helps students solve puzzles that involve discrimination of colors, shapes, and patterns. Students must categorize.
3	9	Students try to discover which chemicals are responsible for which mutations of monsters.
4	8	Students try to discover the numerical rules that allow passage through a king's castle.
2	8	While a small frog hops from lily pad to lily pad, students try to find a pattern in the hops.
4	12	Simulation programs in which students use scientific thinking and reasoning to solve problems. The programs include measurement process, concept development, theory formation, ratio reasoning, and scientific models.
3	8	As they play three different adventures, students learn research skills and check facts in a 100-page reference book (included).
4	8	Students test model cars to see the effects of changing variables.

TITLE	PUBLISHER	DATE	COST	MGT	MDFY
Utility					
APPLEWORKS, SPREADSHEETS FOR MATHEMATICS AND SCIENCE	MECC	1988	$59.00		
AWARD MAKER PLUS	Baudville	1986	$49.95		
BANK STREET SCHOOL FILER	Bank Street School of Education (Sunburst)	1986	$99.00		
CREATE-A-TEST	Create-A-Test	1987	$89.95	Mgt	Mdfy
PC: SOLVE	Pacific Crest Software				
SCIENCE DEPARTMENT	Condor	1988	$59.00		

LOW	HIGH	COMMENTS
6	12	Helps develop analytical skills and higher-order thinking skills.
4	Adult	Allows the creation of certificates of professional quality with personalized messages for any occasion.
5	Adult	A powerful database management program especially designed to help teachers integrate databases with classroom instruction.
Adult	Adult	A kit with a manual, test construction program, and utility file. Teachers can create multiple-choice, matching, true/false, essay, and other tests.
9	Adult	A problem-solving language with an easy user interface.
Adult	Adult	A program to keep inventories of such items as science equipment, chemicals, and supplies.

ANNOTATED BIBLIOGRAPHY

- Abraham, Michael R., and John W. Renner. "The Sequence of Learning Cycle Activities in High School Chemistry." *Journal of Research in Science Teaching* 23 (1986): 121–143.

  This article describes two chemistry lessons developed using the learning cycle approach as the optimum sequence for achievement of content knowledge.

- Abruscato, Joseph. *Children, Computers, and Science Teaching: Butterflies and Bytes.* Englewood Cliffs, N.J.: Prentice-Hall, 1986.

  This book presents an engaging view of enriching the curriculum with microcomputers, software, hardware, and computer languages. It also discusses such topics as strategies for using one computer in the classroom and what new technologies can offer for science education.

- AETS Ad Hoc Committee on Computers in Science Teaching. "Computers in Science Education: An AETS Position Paper." *Journal of Computers in Mathematics and Science Teaching* 4 (Summer 1985): 17–20.

  This report describes the roles of computers in science education and discusses alternatives for enhancing computer knowledge of pre-service and in-service teachers.

- Anderson, Charles W. "Strategic Teaching in Science." In B. F. Jones, A. S. Palinscar, D. S. Ogle, and E. G. Carr (Eds.), *Strategic Teaching and Learning: Cognitive Instruction in the Content Areas.* Alexandria, Va.: ASCD, 1987.

  This book discusses "strategic teaching" in several curriculum areas. This chapter focuses on helping students develop process skills through strategic teaching in science.

- Baird, William E. "Status of Use: Microcomputers in Science Teaching." *Journal of Computers in Mathematics and Science Teaching* 8 (Summer 1989): 14–25.

  This article discusses historical, current, and future trends. At the present time, very few science teachers are using computers.

- Bitter, Gary. "CD-ROM Technology and the Classroom of the Future." *Computers in the Schools* 5(1/2) (1988): 23–34.

  This article presents some good definitions, descriptions, and applications of CD-ROM.

- Brasell, Heather. "The Effect of Real Time Laboratory Graphing on Learning Graphic Representations of Distance and Velocity." *Journal of Research in Science Teaching* 24 (1987): 385–395.

This article describes the use of MBLs in physics classes and the greater learning experienced by students who watch graphs being created in real time.

- Dalton, David W. "The Effects of Cooperative Learning Strategies on Achievement and Attitudes During Interactive Video." *Journal of Computer-Based Instruction* 17(1) (1990): 8–16.

  This article explores the effects on learner achievement, attitudes, and interaction produced by cooperative use of interactive video science lessons.

- Driver, Rosalind. *The Pupil as Scientist.* Milton Keynes, England: Open University Press, 1983.

  This book offers an excellent introduction to how to teach elementary science through the understanding of the thinking of young students. Science teachers must recognize and act on preconceptions and alternative frameworks. It is applicable to all grade levels.

- Ellis, James D. (Ed.) *Information Technology and Science Education*, 1988 AETS Yearbook. Columbus, Ohio: ERIC, 1989.

  This yearbook, written by leaders in the field of computers in science education, contains chapters on MBLs, telecommunications, optical technologies, status of microcomputers in the science classroom, general uses and effects of technology, cooperative learning, VOYAGE OF THE MIMI, and teacher preparation.

- Friedler, Yael, Rafi Nachmias, and Marcia C. Linn. "Learning Scientific Reasoning Skills in Microcomputer-Based Laboratories." *Journal of Research in Science Teaching* 27 (1990): 173–191.

  An MBL curriculum, called the Computer as Lab Partner, was designed to develop students' observation, prediction, and scientific reasoning skills about heat and temperature problems.

- Good, Ron. "Artificial Intelligence and Science Education." *Journal of Research in Science Teaching* 24 (1987): 325–342.

  This article presents a discussion of ICAI and effective instruction in science education, including several software packages that have been developed. A modified learning cycle is also proposed.

- Harlen, Wynne (Ed.). *Primary Science: Taking the Plunge.* London: Heinemann, 1985.

  This book discusses how to teach elementary school science most effectively. The techniques are also applicable to middle and high school.

- Leonard, William H. "A Comparison of Student Reactions to Biology Instruction by Interactive Videodisc or Conventional Laboratory." *Journal of Research in Science Teaching* 26 (1989): 95–104.

  This article reports on a research study showing that a videodisc provides enrichment but not replacement for a "wet" lab.

- Linn, Marcia. "An Apple a Day." *Science and Children* 25(3) (1987): 15–18.

  This article contains a description of the use of an MBL heat and temperature unit with middle school students in California.

- Mokros, Janice R., and Robert F. Tinker. "The Impact of Microcomputer-Based Labs on Children's Ability to Interpret Graphs." *Journal of Research in Science Teaching* 24 (1987): 369–383.

  Middle school students can learn to communicate with graphs in the context of an appropriate MBL. An MBL uses multiple modalities, pairs the real-time event with its graphical representation, and eliminates the drudgery in graph production.

- Nachmis, Rafi, and Marcia C. Linn. "Evaluations of Science Laboratory Data: The Role of Computer-Presented Information." *Journal of Research in Science Teaching* 24 (1987): 491–506.

  This article examines how students evelute information acquired in the science laboratory using MBLs, probes, and the tables and graphs presented by the computer.

- Osborne, Roger, and Peter Freyberg. *Learning in Science: The Implications of Children's Science*. London: Heinemann, 1985.

  This book describes children's beliefs about science concepts and discusses the ways in which children learn science.

- Papert, S. *Mindstorms: Children, Computers, and Powerful Ideas*. New York: Basic Books, 1980.

  This book is the classical introduction by the "father of Logo" to the use of computers for creating "microworlds."

- Pogge, Alfred F., and Vincent N. Lunetta. "Spreadsheets Answer 'What If ...?'" *The Science Teacher* 54(8) (November 1987): 46–49.

  This article gives several examples of spreadsheets for population and ecology studies and pendulum problems. The computer does the calculations, freeing students to question and analyze the problems.

- Rakow, Steven J. *Teaching Science as Inquiry*. PDK Fastback #246. Bloomington, Ind.: Phi Delta Kappa, 1986.

  This booklet discusses the inquiry approach as the best way to teach and learn science.

- Rakow, Steven J., and Terry R. Brandhorst. *Using Microcomputers for Teaching Science*. PDK Fastback #297. Bloomington, Ind.: Phi Delta Kappa, 1989.

  Topics include the computer environment, computer-assisted learning, simulations, laboratories, and emerging technologies.

- Renner, John, Michael Abraham, and Howard Birnie. "The Necessity of Each Phase of the Learning Cycle in Teaching High School Physics." *Journal of Research in Science Teaching* 25 (1988): 39–58.

  The three phases of the learning cycle are distinct and each is necessary. The philosophy is presented with several well-developed examples of a physics lesson developed in this format.

- Renner, John W., and Edmund A. Marek. *The Learning Cycle and Elementary School Science Teaching*. Portsmouth, N.H.: Heinemann, 1988.

  This book provides an explanation of each step of the learning cycle and how to use it to teach elementary science.

- Rivers, Robert, and Edward L. Vockell. "Computer Simulations to Stimulate Scientific Problem Solving." *Journal of Research in Science Teaching* 24 (1987): 403–415.

  This article reports on the use of computerized simulations, which enable students to master the objectives of a biology course and to learn important, generalized problem-solving techniques.

- Sherwood, Robert D. "Optical Technologies: Current Status and Possible Directions for Science Instruction." In J. D. Ellis (Ed.), *Information Technology and Science Education*, 1988 AETS Yearbook. Columbus, Ohio: ERIC, 1989.

  This chapter discusses such questions as, What are the major hardware and software technologies currently in use in videodiscs and CD-ROM? Why might their use provide a learning environment especially useful for science instruction?

- Shymansky, James A., Larry V. Hedges, and George Woodworth. "A Reassessment of the Effects of Inquiry-Based Science Curricula of the 60's on Student Performance." *Journal of Research in Science Teaching* 27 (1990): 127–144.

Inquiry-based curricula of the 1960s and 1970s were more effective in enhancing student performance than were traditional textbook-based programs of the time.

- Strickland, A. W., and T. Hoffer. "Databases, Problem-Solving and Laboratory Experiences." *Journal of Computers in Mathematics and Science Teaching* 9(1) (1989): 19–28.

    This article gives a description of a problem-solving activity integrating a laboratory investigation of the physical properties of materials with a database.

- Vernier, David. *How to Build a Better Mousetrap and 13 Other Science Projects Using the Apple II.* Portland, Or.: Vernier Software, 1986.

    This booklet describes fourteen projects that can be built to interface to the computer and describes experiments that can be done with them. All supplies are available at local electronic stores or from Vernier Software.

- Vockell, Edward L., and Eileen Schwartz. *The Computer in the Classroom.* Watsonville, Calif.: Mitchell, 1988.

    This book provides a good introduction to the whole range of instructional applications of the microcomputer to education. It is a useful tool for training teachers to use computers more effectively.

- Vockell, Edward L., and Robert van Deusen. *The Computer and Higher-Order Thinking Skills.* Watsonville, Calif.: Mitchell, 1989.

    This book, a companion volume to the present book, focuses in detail on using the computer to teach higher-order thinking skills in all curriculum areas. It advocates a use of guided discovery with a focus on generalizing skills to new areas.

- Woerner, Janet. "The Apple II Microcomputer as a Laboratory Tool." *Journal of Computers in Mathematics and Science Teaching* 7(1–2) (1987–1988): 34–37, 43.

    Discussion of the building of a simple I/O interface for the Apple II series and science experiments which can be done with it.

VENDORS OF SCIENCE SOFTWARE

Accu-Weather, Inc.
619 W. College Avenue
State College, PA 16801

Advanced Ideas
2902 San Pablo Avenue
Berkeley, CA 94702

Agency for Instructional
Technology
Box A
Bloomington, IN 47402

American Only, Inc.
13361 Frati Lane
Sebastopol, CA 95472

Baudville
1001 Medical Park Drive
Grand Rapids, MI 49506

Beagle Brothers
3990 Old Town Avenue,
Suite 102C
San Diego, CA 92110

Bergwall Educational Software
106 Charles Lindbergh Blvd.
Uniondale, NY 11553

Biolearning Systems, Inc.
420 Lexington Avenue, Suite 2735
New York, NY 10017

Broderbund Software
17 Paul Drive
San Rafael, CA 94903–2101

Carina Software
830 Williams Street
San Leandro, CA 94577

CDL: Cambridge Development
Laboratories, Inc.
42 Fourth Avenue
Waltham, MA 02154

Claris
440 Clyde Avenue
Mountain View, CA 94043

COMPress
PO Box 102
Wentworth, NH 03282

Condor Computing
7255 Stewart Road
Dane, WI 53529

Conduit
University of Iowa
Oakdale Campus
Iowa City, IA 52242

Create-A-Test
80 Tilley Drive
Scarborough, Ontario, Canada
MIC 2G4

Cross Educational Software
1802 N. Trenton Street
PO Box 1009
Ruston, LA 71270

Cygnus Software
8002 E. Culver
Mesa, AZ 85207

D. C. Heath
Collamore Educational Publishing
125 Spring Street
Lexington, MA 02173

Data Command
PO Box 548
Kankakee, IL 60901

Datatech Software Systems
19312 E. Eldorado Drive
Aurora, CO 80013

Decision Development
Corporation
2680 Bishop Drive, Suite 122
San Ramon, CA 94583

DesignWare
185 Berry Street
San Francisco, CA 94114

Diversified Educational
Enterprises
725 Main Street
Lafayette, IN 47901

E & L Software
95 Richardson Road
Chelmsford, MA 01863

Earthware Computer Services
PO Box 30039
Eugene, OR 97403

Educational Activities
PO Box 392
Freeport, NY 11520

Educational Software Products
12 Bella Vista Place
Iowa City, IA 52240

EduTech
1927 Culver Road
Rochester, NY 14609

EME
Old Mill Plain Road
PO Box 2805
Danbury, CT 06813–2805

EMF
Educational Materials and
Equipment Co.
PO Box 17
Pelham, NY 10803

Estes Industries
1295 H Street
Penrose, CO 81240

Focus Media
839 Stewart Avenue
PO Box 865
Garden City, NY 11530

Grolier Educational Publishing
95 Madison Avenue
New York, NY 10016

Hartley Courseware
123 Bridge
Dimondale, MI 48821

High Technology Software Products
8200 N. Classen Blvd., Suite 104
Oklahoma City, OK 73114

HRM Software, A Division of Queue
338 Commerce Drive
Fairfield, CT 06430

IBM
PO Box 1329
Boca Raton, FL 33432

Instructivision, Inc.
3 Regent Street
Livingston, NJ 07039

Intellimation Library for the Mac
Department XA
130 Cremora Drive
PO Box 1922
Santa Barbara, CA 93116

Interstel Corporation
17317 El Camino Real
Houston, TX 77058

Interstellar
4921 Mackleman Drive
Oklahoma City, OK 73135

Island Software
Box 300, Dept. B
Lake Grove, NY 11755

J & S Software
135 Haven Avenue
Port Washington, NY 11050

Kemtec Educational Corporation
9889 Crescent Park Drive
West Chester, OH 45069

The Learning Company
545 Middlefield Road
Menlo Park, CA 94025

Macmillan McGraw-Hill
220 E. Danieldale Road
De Soto, TX 75115

Mark Davids
21825 O'Connor
St. Clair Shores, MI 48080

Marshware
PO Box 8082
Shawnee Mission, KS 66208

MECC: Minnesota Educational
Computing Consortium
3490 Lexington Avenue North
St. Paul, MN 55126

Michigan State University
Instructional Media Center
Marketing Division
East Lansing, MI 48823–0610

Microcomputer Workshops Corp.
225 Westchester Avenue
Portchester, NY 10573

Milliken Publishing Company
PO Box 21579
St. Louis, MO 63132

National Dairy Council
6300 N. River Road
Rosemont, IL 60018–4233

National Geographic Society
Educational Services
Washington, DC 20036

Optical Data Corporation
30 Technology Drive
Warren, NJ 07060

Pacific Crest Software
887 NW Grant Avenue
Corvallis, OR 97330

PC-SIG
1030-D E. Duane Avenue
Sunnyvale, CA 94086

Petrocci Freelance Associates
652 N. Houghton Road
Tucson, AZ 85748

Prentice-Hall
Route 59 at Brook Hill Drive
West Nyack, NY 10955

Profiles, Inc.
507 Highland Avenue
Iowa City, IA 52240

Project SERAPHIM
Department of Chemistry
University of Wisconsin,
Madison
Madison, WI 53706

Rand McNally
8255 N. Central Park
Skokie, IL 60076

ROF: Richard O. Fee
PO Box 216
Keysport, IL 62253

Sargent-Welch
7350 North Lender Avenue
PO Box 1026
Skokie, IL 60076–1026

Scholastic
730 Broadway
New York, NY 10003

Scott Resources, Inc.
ESNR Division
PO Box 2121
401 Hickory Street
Ft. Collins, CO 80522

Sensible Software
335 E. Big Beaver, Suite 207
Troy, MI 48083

Software Publishing Corporation
1901 Landings Drive
Mountain View, CA 94043

Spectrum Holobyte
2061 Challenger Drive
Alameda, CA 94501

Sublogic Communications
713 Edgebrook Drive
Champaign, IL 61820

Sunburst Communications
39 Washington Avenue
Pleasantville, NY 10570

Synergistic
830 N. Riverside Drive
Renton, WA 98055

Teachers Idea & Information
Exchange
PO Box 6229
Lincoln, NE 68506

Thornton Associates, Inc.
1432 Main Street
Waltham, MA 02154

Thorobred Software
PO Box 1131
Murray, KY 42071

Tom Snyder Productions
123 Mt. Auburn Street
Cambridge, MA 02138

Ventura Educational Systems
3440 Brokenhill Street
Newbury Park, CA 91320

Vernier Software
2920 SW 89th Street
Portland, OR 97225

Walt Disney Telecommunications
500 E. Buena Vista Street
Burbank, CA 91521

Ward's Natural Science
Establishment
5100 W. Henrietta Road
PO Box 92912
Rochester, NY 14692–9012

APPENDIX E

SCIENCE LASERDISCS

This appendix lists by category a large number of laserdiscs that can be employed in science education. Indicated grade levels are approximations. Likewise, prices vary; and the prices listed here are likely to have changed.

"Type" refers to constant angular velocity (CAV) and constant linear velocity (CLV). CAV is the standard-play laser videodisc format most commonly used for computer interactive applications. Each frame can be addressed and presented individually within seconds on most players. CLV is the extended-play videodisc format most commonly used for linear applications such as movies, concerts, and the like. Searching is limited to prerecorded chapter stops and minutes and seconds. This format does not easily lend itself to computer interactive applications.

TITLE	PUBLISHER	GRADE
Chemistry		
Chemical Reactions and Solubilities	Falcon Software	12
Chemistry I and II	GPN	11–12
Doing Chemistry	American Chemical Society	9–12
Gases and Air Analysis	Falcon Software	12
Periodic Table Videodisc: Reactions of the Elements	JCE Software	6–12
Rates and Equilibrium	Falcon Software	12
Understanding Chemistry and Energy	Systems Impact	6–12
Earth Science		
Africa's Stolen River (National Geographic)	Image Entertainment	4–12
Bermuda Triangle (NOVA)	Image Entertainment	6–12
Born of Fire (National Geographic)	Image Entertainment	4–12
Earth Science	Systems Impact	5–12
Earth Science Sides 1–2: Geology and Meteorology	Optical Data Corporation	7–12
Earth Science Sides 3–6: Astronomy and the Sun	Optical Data Corporation	7–12
Footsteps of Giants	Image Entertainment	5–12
Gems and Minerals: Smithsonian Laserdisc Collection	Lumivision	3–12
Interactive Science Series: Weather	Health EduTech	4–9
The Living Textbook: Astronomy and the Sun	Optical Data Corporation	7–12
The Living Textbook: Geology and Meteorology	Optical Data Corporation	7–12
NASM: Archival Videodisc 6	Smithsonian Institute Press	6–12
The Night Sky (2nd ed.)/The Universe	Encyclopaedia Britannica Education Co.	6–12
One Small Step (NOVA)	Image Entertainment	6–12
The Planets	National Geographic	7–12
Planetscapes: Space Disc Vol. 2	Optical Data Corporation	3–12
Predictable Disaster (NOVA)	Image Entertainment	6–12
Space Shuttle: Space Disc Vol. 3	Optical Data Corporation	3–12
UFO's Are We Alone? (NOVA)	Image Entertainment	6–12
Volcanoes: Exploring the Restless Earth/ Heartbeat of a Volcano	Encyclopaedia Britannica Education Co.	4–12
Voyager Gallery: Space Disc Vol. 1	Optical Data Corporation	3–12
What Makes Clouds?/What Makes the Wind Blow?	Encyclopaedia Britannica Education Co.	4–12
Windows on Science–Earth Science	Optical Data Corporation	4–6

TYPE	PRICE	DESCRIPTION
CAV	$400.00	Chemical reactions, oxidation/reduction, and redox solubilities
CAV	425.00	Video simulation of titrations and determining unknowns
CAV	(Call)	One hundred twenty-two experiments and demonstrations to encourage hands-on activity
CAV	400.00	Pressure, PVT introduction and experiments, density and molecular weight, air analysis
CAV	50.00	Action and still shots of each element and its common reactions
CAV	400.00	Introduction to rates, equilibrium, and LêChatelier's principle
CAV	1950.00	A 20-lesson minicourse on foundational chemistry
CLV	29.95	The appearance and disappearance of the Savuti Channel in Botswana
CLV	29.95	NOVA takes a hard look at this watery graveyard
CLV	29.95	Record of crustal plate movement, causing quakes and volcanoes
CAV	2600.00	Thirty-one lessons for physical science minicourse, foundational to earth science
CAV	495.00	Broad coverage of all topics in geology and meteorology
CAV	895.00	Two-videodisc set with broad coverage of topics in astronomy
CLV	29.95	Chronicles 25 years of America's adventure in space
CAV	29.95	A journey behind the scenes to the gem and mineral collection
CAV	1860.00	Three-disc set with units on wind, air masses, water cycle, and clouds
CAV	1195.00	Two discs with 15,000 slides, 160 movie clips on basic concepts
CAV	845.00	Seventy-two hundred slides, 34 movie clips on all basic concepts in geology, meteorology
CAV	55.00	Lunar disc with 70,000 images
CAV	99.00	Explores human inquiry into the sun, moon, planets, and stars
CLV	29.95	First successful space launch and moon landing
CAV	87.50	Pictorial database and program on the planets
CAV	295.00	Approximately 12,000 still frames that scan fine details of planets
CLV	29.95	Earthquake disaster, techniques for monitoring and predicting them
CAV	395.00	Documents the testing and flights of the first 11 shuttle missions
CLV	29.95	Explores UFO sightings
CAV	99.00	On-location photography and animated drawings/eruption of Kilauea
CAV	195.00	Slides, diagrams, maps, text and movie clips from Voyager missions
CAV	99.00	Observation of cloud formation, conditions for air movement demonstration
CAV	595.00	Two discs, covers all basic concepts, includes hands-on activities

TITLE	PUBLISHER	GRADE
Elementary Science		
Science Helper K–8 CD-ROM	PC-SIG	K–12
Life Science		
African Wildlife (National Geographic)	Image Entertainment	3–12
All American Bear (NOVA)	Image Entertainment	6–12
Among the Wild Chimpanzees (National Geographic)	Image Entertainment	3–12
Animal Olympians (NOVA)	Image Entertainment	6–12
Animals in Spring and Summer (2nd ed.)/ Autumn and Winter	Encyclopaedia Britannica Education Co.	1–6
Audubon/Galapagos: My Fragile World (National Geographic)	Image Entertainment	4–12
Australia's Improbable Animals (National Geographic)	Image Entertainment	1–12
Bio Libe Encyclopedia	Bio Libe	K–12
Bio Sci Videodisc	Videodiscovery	K–12
Cell Biology: I Motion and Function of the Living Cell	Videodiscovery	6–12
Creatures of the Mangrove (National Geographic)	Image Entertainment	3–12
Creatures of the Namib Desert (National Geographic)	Image Entertainment	3–12
Death Trap	Videodiscovery	6–12
Desert Whales (Cousteau)	Image Entertainment	1–12
Dinosaurs: Fantastic Creatures That Ruled the Earth	Lumivision	4–12
Dragons of Galapagos (Cousteau)	Image Entertainment	3–12
Ears and Hearing (2nd ed.)/Eyes and Seeing	Encyclopaedia Britannica Education Co.	9–12
ECO-Insights	Access Network	7–10
Encyclopedia of Animals: Vol. 1–8	Pioneer Ldca	1–12
Exotic Plants: A Videodisc Compendium	VT Productions	K–12
Flight of the Penguins (Cousteau)	Image Entertainment	3–12
Forgotten Mermaids (Cousteau)	Image Entertainment	3–12
Gorilla (National Geographic)	Image Entertainment	3–12
Great Ape (National Geographic)	Image Entertainment	4–12
Great Whales (National Geographic)	Image Entertainment	3–12
Grizzlies (National Geographic)	Image Entertainment	3–12
Human Brain (2nd ed.)/Nervous System (3rd ed.)	Encyclopaedia Britannica Education Co.	7–12
Incredible Human Machine (National Geographic)	Image Entertainment	4–12

TYPE	PRICE	DESCRIPTION
	$195.00	Contains activities from SCIS, SAPA, USMES, ES, and other science programs
CLV	29.95	Close-ups of birth, death, and survival of wildlife in Namibia
CLV	29.95	The hibernation of the North American black bear
CLV	29.95	Two landmark decades of Jane Goodall's work among the chimpanzees
CLV	29.95	A look at the beauty, power, speed, and endurance of animals
CAV	99.00	Close-up sequences that show animal adaptation to yearly cycles
CLV	29.95	A journey to the almost prehistoric world of the Galapagos Islands
CLV	29.95	A survey of Australia's wide array of curious life forms
CAV	250.00	Over 4,600 color slides of the earth's fauna and flora
CAV	549.00	Six thousand stills, charts, diagrams, and motion pictures on biology
CAV	549.00	Eighty-six films, stills in five categories of cell structure and process
CLV	29.95	Motion pictures of the bizarre creatures of the tidal mangrove
CLV	29.95	Describes the desert's vast wilderness and its creatures
CLV	125.00	A study of the natural history of carnivorous plants
CLV	29.95	A study of the 5,000-mile migration of the California gray whale
CAV	29.95	A behind-the-scenes tour of the Smithsonian's Dinosaur Hall
CLV	29.95	A study of the marine iguana, a primitive reptile
CAV	99.00	Visuals on the structure and function of ear and eyes
CAV	175.00	Introduction to basic ecology in the Canadian Rocky Mountains
CAV	99.95	Eight-disc set on behavior of mammals, birds, reptiles, insects, fish
CAV	225.00	Two thousand photos of tropical, subtropical, and other exotic plants
CLV	29.95	Life of the Antarctic penguin above and below the sea
CLV	29.95	A film on the manatee in the Cypress Glades of east Florida
CLV	29.95	A look at the mountain gorilla of central Africa
CLV	29.95	Fieldwork of Diane Fossey and Birute Galdikas-Brindamour on the ape
CLV	29.95	The anatomy, communication, and migratory patterns of great whales
CLV	29.95	A portrait of the powerful, intelligent, and aggressive grizzly
CAV	99.00	Demonstrates brain and nervous system structure and function
CLV	29.95	An exploration into how and why the human body works

TITLE	PUBLISHER	GRADE
Interactive Science Series: Animals	Health Edutech	4–9
Interactive Science Series: Plants	Health Edutech	4–9
Land of the Tiger (National Geographic)	Image Entertainment	3–12
Life Cycles	Videodiscovery	K–12
Life Science 1–4: Molecular, Cell, Human, Plant, and Animal Biology	Optical Data Corporation	7–12
Life Science 5–6: The Frog	Optical Data Corporation	7–12
Life Science Biology I and II	GPN	9–12
Lions of the African Night (National Geographic)	Image Entertainment	3–12
Living Textbook—Life Science Interactive Multimedia Library	Optical Data Corporation	7–12
Living Textbook—The Frog Interactive Multimedia Library	Optical Data Corporation	7–12
Madagascar (National Geographic)	Image Entertainment	4–12
Meiosis (2nd ed.)/Mitosis (2nd ed.)	Encyclopaedia Britannica Education Co.	7–12
The National Zoo: The Zoo Behind the Zoo	Lumivision	K–12
Octopus–Octopus (Cousteau)	Image Entertainment	4–12
Pollination Biology	Videodiscovery	6–12
Rain Forest (National Geographic)	Image Entertainment	3–12
Realm of the Alligator (National Geographic)	Image Entertainment	4–12
Regulating Body Temperature (2nd ed.)/ Digestive System (2nd ed.)	Encyclopaedia Britannica Education Co.	7–12
Relationships	Syscon Corporation	9–12
Respiratory System (2nd ed.)/Endocrine System (2nd ed.)	Encyclopaedia Britannica Education Co.	9–12
Rocky Mountain Beaver Pond (National Geographic)	Image Entertainment	3–12
Save the Panda (National Geographic)	Image Entertainment	3–12
Sharks (National Geographic)	Image Entertainment	4–12
Signs of the Apes/Song of the Whales (NOVA)	Image Entertainment	6–12
Silent Safari Series I and II	Encyclopaedia Britannica Education Co.	K–9
Sound of the Dolphins (Cousteau)	Image Entertainment	3–12
Videodisc in Science Education	Waterford Institute	9–12
Viruses: What They Are and How They Work/ Bacteria (2nd ed.)	Encyclopaedia Britannica Education Co.	7–12

TYPE	PRICE	DESCRIPTION
CAV	$1360.00	Three-sided set with units on vertebrates and invertebrates
CAV	1360.00	Two-sided disc with units on plant structure, survival, and growth
CLV	29.95	Footage on the behavior of the tiger and other animals
CAV	549.00	Four thousand stills and movies on the life cycles of plants and animals
CAV	895.00	Two-disc set provides broad basic coverage of topics in life science
CAV	195.00	Contains anatomy and physiology of the frog, compared to human
CAV	425.00	A video simulation of respiration, climate, and life database
CLV	29.95	A look at the behavior of lions as they search for prey
CAV	1195.00	Two-disk set with 2,700 slides and 164 movie clips on life science
CAV	495.00	Slides and movies on the anatomy and physiology of the frog
CLV	29.95	A look at the flora and fauna of the sunken forests of this island
CAV	99.00	Microscopic footage, animation, and artwork to show cell division
CAV	29.95	A behind-the-scenes tour of the Smithsonian's living collection
CLV	29.95	A view of the octopus from the Mediterranean to Puget Sound
CAV	495.00	Filmed sequences that deal with all facets of flower pollination
CLV	29.95	A journey to the dense rain forests of Costa Rica
CLV	29.95	A study of the behavior of the alligators of the Okefenokee swamp
CAV	99.00	Physiological and behavioral mechanisms used by animals
CAV	145.00	Instruction investigating the relationships among organisms
CAV	99.00	Live action and animation show structure and function of systems
CLV	29.95	Beavers create a pond that supports a community of organisms
CLV	29.95	Join scientists as they track the panda through bamboo forests
CLV	29.95	Visit the underwater kingdom of the shark
CLV	29.95	Explores language in the animal world
CAV	99.00	Film of mating rites, predation, birth and death of animals
CLV	29.95	A journey to the Straits of Gibraltar to study the dolphin
CAV	199.00	Four units of instructional materials for life science courses
CAV	99.00	Explores viral/bacterial structure, reproduction, and diseases

TITLE	PUBLISHER	GRADE
Whale Watch (NOVA)	Image Entertainment	3–12
Whales (Cousteau)	Image Entertainment	3–12
Whales (National Geographic)	National Geographic	1–12
White Wolf (Cousteau)	Image Entertainment	4–12
Windows on Science—Life Science	Optical Data Corporation	4–6
Work of the Heart (2nd ed.)/Muscles: Structure and Function	Encyclopaedia Britannica Education Co.	7–12

Physical Science

Interactive Science Series: Energy	Health Edutech	4-9
Invisible World (National Geographic)	Image Entertainment	3–12
Living Textbook—Physical Science Interactive Multimedia Library	Optical Data Corporation	7–12
Physical Science Sides 1–4	Optical Data Corporation	7–12
Physics and Automobile Collisions	John Wiley	11–12
Physics I and II	GPN	11–12
Physics of Sports	Videodiscovery	11–12
Puzzles of Tacoma Narrows Bridge	Videodiscovery	9–12
Skylab Physics	American Association of Physics Teachers	11–12
TLTG Physical Science	Texas Learning Technology Group	8–10
Windows on Science—Physical Science	Optical Data Corporation	4–6

TYPE	PRICE	DESCRIPTION
CLV	29.95	Whales' annual migration from California to Alaska and back
CLV	29.95	Search of the finback, sperm, and killer whale in the Indian Ocean
CAV	97.50	Motion picture and stills of the underwater world of whales
CLV	29.95	A film about the behavior of white arctic wolves
CAV	595.00	Two-disc set with hands-on activities covering all basic concepts
CAV	99.00	Observe action of heart and three types of muscle
CAV	1860.00	Three-sided set containing units on energy and forms of energy
CLV	29.95	Events that are too small, large, fast, or slow or not visible
CAV	1195.00	Two-disc set with 2,500 physical science slides and 90 movie clips
CAV	895.00	Two-disc set with broad coverage of all topics in physical science
CAV	170.00	High-speed films of automobile collisions to study momentum
CAV	425.00	Two lessons, presenting motion and motion transformation problems
CAV	300.00	Visual record of 20 athletic events, scientifically analyzed
CAV	159.95	Three levels of analysis for the Tacoma Narrows bridge collapse
CAV	150.00	Twelve Skylab experiments and 400 photographs, with teaching manual
CAV	172.00	One hundred sixty-hour physical science curriculum: one-half chemistry, one-half physics
CAV	795.00	Three-disc set, 11 units of hands-on activities in physical science

INDEX